KU-185-790

A Study of the Application of Statistical Sampling to External Auditing

A Study of the Application of Statistical Sampling to External Auditing

T. W. McRAE BSc(Econ), CA
(Professor of Finance, University of Bradford Management Centre)

The Institute of Chartered Accountants
in England and Wales
Chartered Accountants' Hall, Moorgate Place
London EC2P 2BJ
1982

This book consists of a research study undertaken on behalf of the Research Sub-Committee of The Institute of Chartered Accountants in England and Wales. In publishing this book the Institute considers that it is a worthwhile contribution to discussion, but neither the Institute nor the Research Sub-Committee necessarily shares the views expressed, which are those of the author alone.

No responsibility for loss occasioned to any person acting or refraining from action as a result of any material in this publication can be accepted by the author or publisher.

ISBN 0 85291 315 X

Printed by The Institute of Chartered Accountants in England and Wales

CONTENTS

PART A

The Choice of a Sampling Plan

Page

Page

Page

Page

Page

Page

PART B

A brief study of some other aspects of statistical sampling

Page

Page

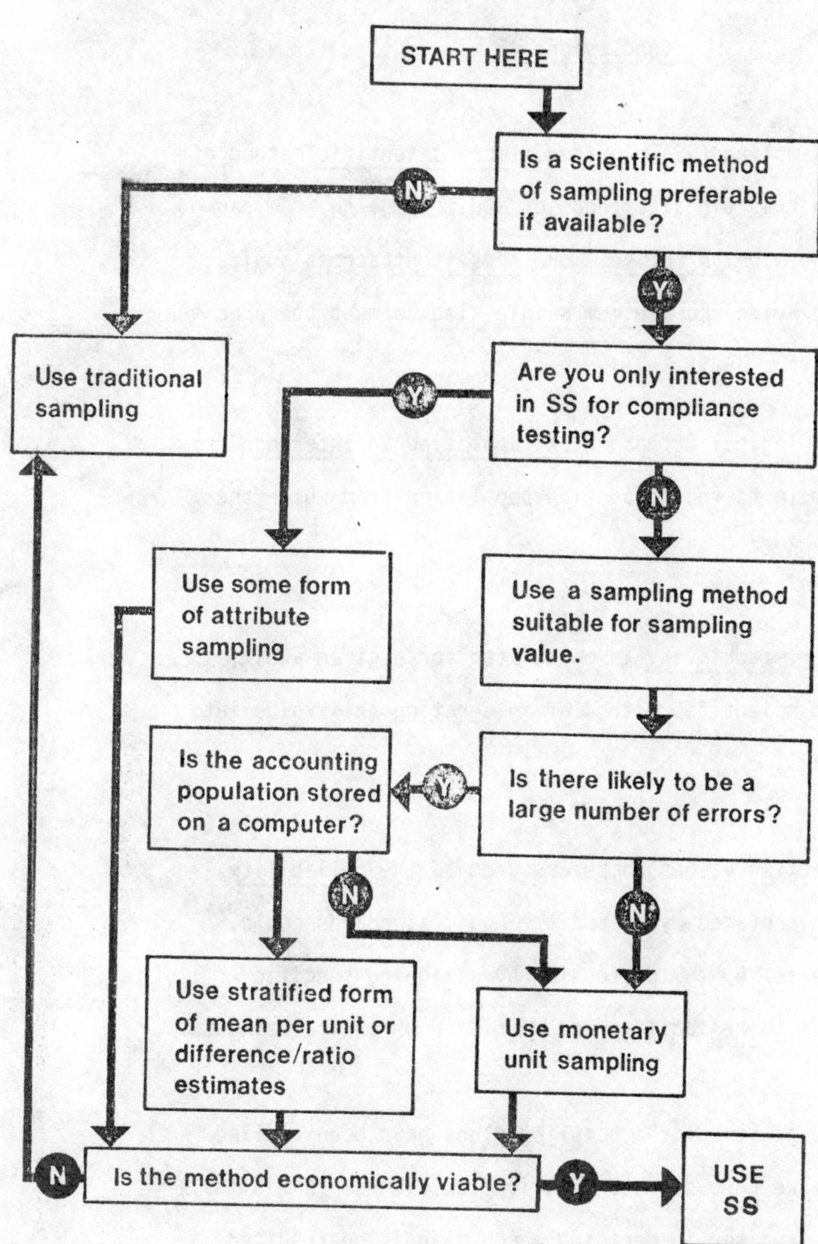

LOGIC DIAGRAM FOR DECIDING WHETHER OR NOT TO USE SS

The uses of statistical sampling in external auditing

Conclusions

1. Statistical sampling (SS) provides a more scientific method of sampling than traditional methods of audit sampling. SS can be shown to be more objective, more consistent across audits, and also to provide the minimum sample size to meet the precise objectives of the audit.

   SS provides a logical method of extrapolating an inference from the audit sample to the accounting population from which the sample is drawn.

2. If a value for materiality is calculated for a given audit, SS provides a scientific method of integrating this value into the audit plan.

3. SS allows specific values to be assigned to the reliability and precision provided by an audit. Audit standards could, in theory, be set by the profession by assigning specific minimum values to reliability and precision on audits using SS.

4. Several types of scientific sampling plans have been applied to auditing. We have devised a logic flow diagram (Exhibit 9.3) to assist the auditor in deciding which plan is best suited to his needs. If the auditor wishes to use SS for estimating error rate, we suggest the use of some form of acceptance sampling. If the auditor wishes to use SS for estimating an

upper limit on error value we suggest the use of monetary unit sampling, so long as the expected error rate is low. If the expected error rate is high we suggest the use of difference estimates with stratification.

5. We believe that the statistical approach entitled the Bayesian method can be used in the context of external auditing.

6. A survey of current usage of SS in external auditing by ICAEW firms found that approximately one third of large accounting firms made extensive use of SS, one third used the method on suitable audits and one third made little or no use of SS in auditing. We noted a trend towards extending the use of the method among large firms.

   Of the order of 10% of medium sized accounting firms (> 4 partners) used SS on suitable audits.

   The usage of SS by AICPA firms in the United States appears to be about double that in the UK.

7. At present courses on SS in the UK would seem to be inadequate except for the in-house courses run by the large accounting firms. Existing SS courses are mainly introductory or of the appreciation type. More practical courses are needed. Several admirable audio-visual and programmed courses are available in the USA and Canada.

   An extensive literature of both books and articles is available on SS but much of this is too theoretical, more practical literature is needed.

We estimate that a programme of 8 hours is needed to teach the application of SS to compliance testing. At least 24 hours is needed to teach the SS methods suited to substantive testing of value.

8. The increased use of the digital computer for storing and processing accounting information is likely to have a major impact on the use of SS in auditing. This trend is already evident in California. The computer can perform SS sampling and evaluation at an economic cost. Several SS software packages are currently available.

In the future the computer may be used to analyse error patterns to select the most economical form of SS for this audit. The computer may also be needed to calculate the upper bound on error value. The computer is also needed when stratification of the accounting population is called for.

If, in the future, SS becomes a generally accepted audit technique the computer will have made it so.

9. We found no evidence to suggest that a statistically validated audit sample will carry more evidential weight in a court of law compared to a traditional audit sample. There is some evidence from the United States to suggest that the Law may be beginning to weigh up the adequacy of audit samples. This could lead towards statistical evaluation of audit sample size.

10. Since auditing is concerned with detecting errors we searched for surveys of audit error type and frequency. We located only one such survey, this was a recent study of error frequency in audits by a leading accounting firm in the United States. So far as we

can discover no audit firm, anywhere in the world, maintains a central record of type and frequency of errors discovered in audits. Such a record would be an invaluable aid to improving sampling plans in auditing.

We conclude that SS does not conflict with an auditor's judgement. SS allows an auditor's judgement to be concentrated on those areas of the audit where it is most needed, that is in those areas where statistical inference cannot be applied. SS simply ensures that the sampling process is being handled in a logical economical and consistent way.

Recommendations

1. Since statistical sampling improves the quality of audit we recommend that action should be taken to encourage its use.

2. SS is best suited to large and medium sized (> 4 partners) accounting firms auditing limited liability companies. We recommend that, in the first place, efforts to encourage the use of SS in auditing ought to be concentrated in this area.

3. We do not recommend that mandatory standards should be set for reliability and precision in audits. However we consider that it would be helpful to firms considering the adoption of SS if the standards currently employed in SS were generally known (i.e. confidence levels and precision limits).

4. Several methods of scientific sampling are currently used in auditing. We recommend that some form of acceptance sampling is best suited to compliance testing and some form of monetary unit sampling best suited to substantive testing of value. We caution, however, that where error rates tend to be relatively high, as with inventory, MUS may not be suitable and some type of auxiliary estimator, such as difference estimate with stratification, should be used.

5. The training facilities in SS are unsatisfactory except within the largest accounting firms. We make the following recommendations.

   (a) A series of courses in SS should be made available. In particular a course should be available on the practical implementation of SS and a seminar on current problems in applying SS, for practitioners.

(b) The excellent audio-visual courses and training manuals in SS developed in the United States and Canada ought to be adapted and made available in the UK.

(c) The large accounting firms might consider opening their in-house courses on SS to staff from the medium sized firms. (One large accounting firm already provides this facility)

6. The various accounting institutes might consider providing the following facilities for their members.

(a) A list of statisticians competent to give advice on the application of statistical method to auditing.

(b) A list of computer audit packages incorporating SS which are available to accounting firms.

(c) In the future a central service facility for medium sized firms may be needed to handle sophisticated methods of computer audit, including SS. This central service facility could be set up by the Accounting Institutes.

7. Research on the type, frequency and distribution of errors in accounting populations ought to be encouraged and supported. Individual audit firms ought to set up and maintain a central register of errors discovered and analyse these on a regular basis. This activity will improve the efficiency of the sampling process in auditing.

ACKNOWLEDGEMENTS

I wish to thank the staff of the many accounting firms in the United Kingdom, the United States and Canada who have given so freely of their time in describing their use of statistical sampling in auditing.

I wish particularly to thank the following individuals in the United States and Canada who provided much useful information on the use of SS in those countries.

Abe Akresh, Rod Anderson, Steve Bussey, Bob Elliott, Duane Frisbie, Fred Gysi, Fred Hancock, Abe Frieger, George Lasky, Don Leslie, John Lindsay, John Neter, Maurice Newman, David Solomons, Ken Stringer, Ralph Teitlebaum and George Zuber.

I would also like to thank the staffs of the ICAEW, the AICPA and the CICA for the contracts and source material supplied by them.

T.W.M.

"If the survey (of external auditing practice) is a fair indication of current practice, the state of the art is indeed one of diversity and uncertainty."

"Extent of Audit Testing." CICA 1980.

"Mathematics, a veritable sorcerer in our computerised society. While assisting the trier of fact in the search for truth, must not cast a spell over him."

People v Collins (1968) 438 P2(d) 33
US Circuit.

"When the determination of the individual items to be included in the samples includes personal judgement; one cannot have an <u>objective</u> measure of the reliability of the sample results."

Hansen M.H., Hurwitz W.N. and Madow W.G. "Sample Survey Methods and Theory." Wiley 1964.

Preface

The Technical and Research Committee of the ICAEW set up four research projects to study certain key aspects of external auditing.

These research projects are focused on materiality, sampling, analytical review and fraud. This study represents the findings of the second of these projects.

The study concentrates on the application of statistical methods to audit sampling, but in so doing it inevitably examines some aspects of the traditional methods of audit sampling.

The researcher was asked to study the application and the literature on statistical sampling (SS) with a view to answering **four** questions.

(1) Is it desirable to apply SS to audit work? In other words "What benefits are to be gained by changing from traditional methods of sampling?"

(2) Is SS viable? In other words are the inferences derived from statistical audit samples accurate?

(3) Is SS economic? In other words can SS be implemented at a cost which is not out of line with the costs of traditional sampling?

(4) Among the various techniques of SS currently being applied, which are the most efficient methods available?

We have attempted to answer these questions in such a way that our answers can be understood by accountants with no knowledge of the mathematics of statistics.

There has been much discussion in recent years about the objectives and efficiency of the external audit process. These factors are related. The efficiency of a process cannot be measured until the objectives of that process are clearly identified.

It is possible that SS may allow the objectives of at least one part of the audit process, namely sampling, to be identified and measured both objectively and consistently.

If the profession accepts this conclusion, it surely follows that some consideration should be given to encouraging the use of the technique?

Abbreviations used in the study.

AICPA	American Institute of Certified Public Accountants.
CICA	Canadian Institute of Chartered Accountants.
CIPFA	Chartered Institute of Public Finance and Accounting.
ICAEW	Institute of Chartered Accountants in England and Wales.
ICAS	Institute of Chartered Accountants of Scotland.
NIVRA	Netherlands Institute of Accountants.
SS	Statistical sampling.

Chapter 1.

THE PROBLEM

1.1 Introduction

Under the 1968 Companies Act, which modified the 1948 Act, the accounts of a company with limited liability must be audited at least once a year by a qualified auditor, the external auditor.

The external auditor is required to report on the accounts under audit. This report must state that the balance sheet and profit and loss account have (a) been prepared in accordance with the provision of the various Companies Acts and (b) give a true and fair view of the state of affairs and the profit or loss of the company, for the period under audit. (Sec 14 (3)).

> "The auditors duty is to examine the original books of account kept by the company, to discover any inaccuracies or omissions therein, to examine the company's annual balance sheet and profit and loss account to ensure that they agree with the original books of account (or corrected), and to report on the original books of account and the annual account to the members." (Pennington p. 632)

These requirements present the external auditor with a formidable undertaking. It is not uncommon for the accounts under audit to be made up from several million individual transactions. An auditor cannot be expected to check every one of these transactions.

The auditor has tackled this problem from several different angles. One approach has been to select a sample from the total population of underlying accounting transactions, to check this sample, and to base his subsequent opinion on the quality of the total population of transactions under audit on the results of this sample test.

It is rare to find an audit which does not employ some form of sampling. Sampling has become an important part of the auditors tool kit. However, the validity of conclusions based on audit samples has become a matter of some controversy in recent years. The attack on traditional methods of audit sampling has come from several quarters.

First, it has been claimed that no proper conceptual model of audit sampling has been developed. What exactly is the objective of external audit sampling and how can an auditor or his client ensure that this objective has been achieved? What are the constraints on achieving the given objective in external audit sampling and how can an auditor trade off the value of achieving his audit objective more precisely against the increasing cost of reducing these constraints? How can the conclusions from an audit sample be represented so that they can be objectively assessed by a third party?

Second it has been claimed that the extent of audit testing is not consistent across audits or between auditors. The CICA (1980) study of audit testing, for example, found a wide variance between audits and auditors. Sneed (1979) found that different audit firms demonstrate significantly different degrees of conservatism with regard to audit testing. Sneed considered that the inconsistency between different audit "firms testing policies (were) sufficiently large to (influence) the fair presentation and comparability of financial statements." (p. 95).

Statistical sampling provides a mechanism for reducing this inconsistency between audits and auditors.

A further attack has been mounted on traditional sampling methods by those who claim that traditional methods of audit sampling are inefficient. Inefficient is defined as meaning that (a) <u>more information</u> could be extracted from a sample of the same size if it were drawn in a different way or (b) a smaller sample could extract the same amount of information if the smaller sample were drawn in a different way.

In other words it is claimed that external audit sampling could be carried out in a more economical way.

All of these criticisms can be subsumed within the statement that traditional methods of audit sampling are claimed to be unscientific. Traditional methods fail to provide objective measures of evidence to prove that audit objectives are met, to prove that testing is consistent, and to prove that the method is cost effective.

The claim that traditional methods of audit sampling are not scientific was first articulated by L.L. Vance (1950) in his admirable book. Many similar claims have been made over the following thirty years although few have rivalled the eloquence and economy of the original presentation.

The obvious solution is to adapt the scientific sampling methods developed by statisticians to external auditing.

Over the last thirty years a great deal of work has been devoted to this end. Many scientific audit sampling methods have been developed and subjected to keen analytical scrutiny. Several international accounting firms use statistical sampling (SS) on a regular basis in their auditing procedures.

Despite all of this considerable effort statistical sampling is not yet generally accepted as the standard method of audit sampling in the accounting profession. Various constraints have hampered the development of a universally acceptable scheme for applying statistical methods to audit sampling.

Why is this so? What are these obstacles which have prevented the accounting profession from adopting a scientific scheme of audit sampling?

1.2 Constraints on the adoption of a scientific sampling method by the accounting profession.

The constraints originate from three sources. First from the nature of accounting populations themselves, second from the limitations of the scientific sampling methods applied to audit work and third from the limitations of existing audit procedures and auditors themselves.

A large part of this study will be given over to a discussion of these limitations but at this early stage let us briefly review some of them.

1.2.1 The nature of accounting populations.

The distribution of value in accounting populations is highly skewed. This can make sampling either complicated, expensive or both.

Accounting errors, particularly value errors, are rare, thus as noted in the next section, rather sophisticated statistical techniques have had to be developed to estimate bounds on error value.

The distribution of error values are also usually highly skewed.

1.2.2 Scientific sampling methods.

Since accounting errors are rare and value populations highly skewed the methods of scientific sampling available in standard textbooks have proved of limited use. Auditors have been forced to develop new methods of scientific sampling suited to audit work.

Many conventional methods of scientific sampling which are widely used by statisticians have been shown to be somewhat innaccurate when applied to the peculiar nature of accounting populations. This has been demonstrated by Neter and Loebbecke (1975) and others.

The audit opinion is built up from many interlinked sources. It is difficult to combine these sources into a single quantitative index of confidence. Much work is currently in progress to solve this problem but a definitive solution is not yet available. The confidence in the sampling inference can be measured scientifically but it cannot, as yet, be linked into an objective measure of total assurance.

The size of sample in statistical sampling is determined objectively once the auditor has decided subjectively on the level of confidence and precision he requires in his inference from the audit sample.

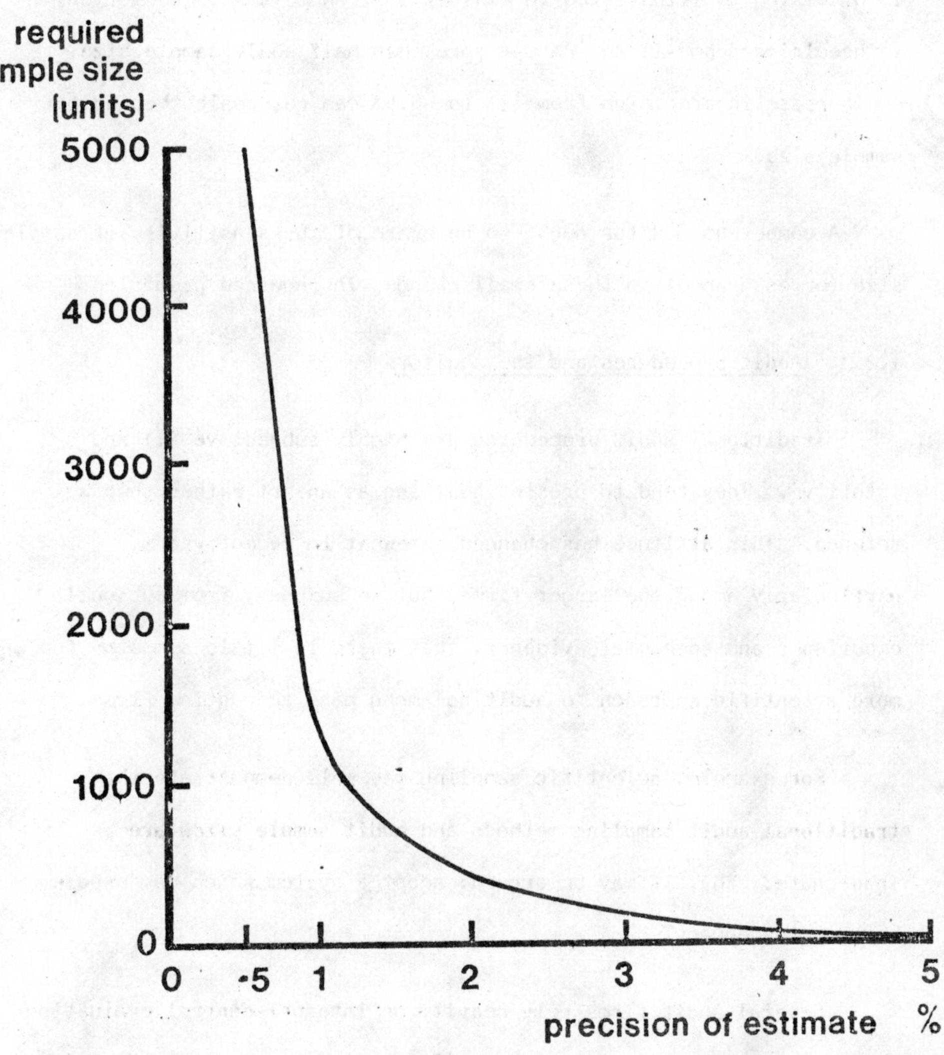

Exhibit 1.1 The diagram illustrates the sensitivity of audit sample size to small changes in the required precision of the estimate - all at a fixed level of confidence and very large population.

The sensitivity of the audit sample size to small variations in precision is illustrated in exhibit 1.1. Note that a reduction in precision from ±1% to ±2% can more than half audit sample size. An increase in precision from ±1% to ±0.5% can quadruple the audit sample size.

A competent auditor ought to be aware of the sensitivity of sample size (or assurance) to these small changes in required precision.

1.2.3 Audit procedures and the auditor.

Traditional audit procedures are highly subjective (1) and intuitive. They tend to present auditing as an art rather than a science. This attitude has changed somewhat in recent years, particularly among the larger firms, but we suspect, from personal experience and anecdotal evidence, that there is little sympathy for a more scientific approach to auditing among many accounting firms.

For example, scientific sampling may well demonstrate that traditional audit sampling methods and audit sample sizes are inadequate. Why, it may be argued, adopt a system which may expose our own inadequacy?

Several audit firms rely heavily on internal control evaluation when seeking assurance about the validity of the accounts under audit. Substantive sampling may either be zero, or very small "walk through" samples may be utilised which make no pretence to be representative. Accounting firms using this philosophy will show no interest in scientific sample evaluation of the accounts under audit.

A variant of the above occurs when the audit fee is under great pressure from the client. The auditor may plead that he cannot afford to draw audit samples large enough to satisfy proper statistical evaluation procedures.

Finally an auditor may, quite legitimately complain that, at the present time, a statistically validated audit sample appears to carry no greater evidential weight than a traditional audit sample in an English law court. (See Chapter 14)

Finally, we come to the limitations imposed by the auditor himself. The main limitation imposed by the auditor is his limited knowledge of statistical theory. This defect is being slowly remedied by the professional accounting examinations but few auditors enjoy sufficient knowledge of statistical theory to discuss the more advanced aspects of SS.

Auditors need not understand a sampling method to use it. But it helps. If auditors do not understand the theory behind a sampling method, they may use it incorrectly, they will certainly feel insecure in its use.

For all of these and other lesser reasons the technique of statistical sampling has been slow to gain acceptance among professional auditors.

1.3 The impact of the computer.

The increasing use of the computer for storing and processing accounting information is likely to accelerate the use of statistical methods in auditing.

Many of the more difficult or rather tiresome chores associated with SS can be delegated to the computer.

Audit programmes including SS packages are already available to handle a wide range of SS techniques.

The computer can be used to calculate the required sample size, select the sample and evaluate the sample results. This leaves the single step of auditing the sample to the auditor.

Statistical sampling affects the methods by which an audit sample is selected and the method by which the sample results are evaluated but SS does not, as yet, alter the traditional methods of auditing the units selected for test. (2)

Since statistical sampling methods are likely to become more sophisticated in the years ahead the computer will play an increasingly important role in audit sampling and sample evaluation.

1.4 The objective of the study.

The objective of this study is to examine the viability of applying scientific sampling methods to external auditing.

We will examine the various SS methods which have been applied to external auditing to identify the strengths and weaknesses of each method.

We will attempt to derive a set of criteria by which the efficiency of an audit sampling method can be judged and we will suggest a scheme for selecting a sampling method suited to both the specific needs of the auditor and the accounting population under audit.

1.5 Organisation of the study.

Chapter two will examine the criteria for assessing the efficiency of an external audit sample. Such criteria cannot be decided without first placing sampling within the overall framework of the external audit process. We will examine the role of sampling in auditing and its relationship to other testing techniques.

Chapter three will review the limited evidence available on the pattern of error discovered in accounting populations. This topic is of some importance when it comes to assessing the suitability of various sampling methods to audit work.

Chapter four will describe the various sampling methods which have been applied to audit work. The statistical formula for calculating sample size by each of these methods is provided.

Chapter five examines the history of applying SS to external audit work. The key ideas are identified and the historical development of SS described.

Chapter six reviews several surveys on the extent of use of SS by external auditors in several countries. The results of a survey of the extent of use of SS in England and Wales by large and medium sized firms is displayed and discussed.

Chapter seven provides a brief discussion of some aspects of the economics of audit sampling. Particular attention is paid to the incremental cost of increasing the sample size.

Chapter eight provides a critical assessment of each of the sampling methods described in chapter four. The merits and limitations of each method are set down, and the viability of each method, under various conditions, is discussed.

Chapter nine suggests a scheme for selecting the most suitable audit sampling method given various prior conditions.

Chapter ten discusses the various options open to an auditor if the audit sample size suggested by SS is significantly in excess of the sample size he would have drawn using traditional sampling methods.

Chapter eleven describes the organisational framework required to apply statistical sampling to external audit.

Chapter twelve reviews the facilities available for training accountants in the use of SS.

Chapter thirteen briefly discusses the contribution that the computer can make towards implementing SS.

Chapter fourteen examines some legal implications of SS, with particular emphasis on the evidential value of a statistically derived audit sample.

Finally chapter fifteen reviews the literature on SS and provides a selective bibliography.

Notes on Chapter One.

1. By subjective we mean that a formal model of audit objectives and constraints is not available for inspection by, say, the client of the auditor or a third party. Note the wide disparity in audit procedures on identical sets of accounts noted in the CICA (1980) study, "Extent of Audit Testing".
2. The computer is being used to test the "reasonableness" of accounting values selected for test i.e. "odd" items sampled can be subjected to a more detailed scrutiny. This approach represents "key item" sampling not representative sampling.

Chapter 2.

CRITERIA FOR ASSESSING THE EFFICIENCY OF AN EXTERNAL AUDIT SAMPLE.

2.1 Introduction

We wish to examine the proposition that a statistically derived external audit sample is more efficient than an audit sample derived by traditional methods.

In order to test this hypothesis we need to set up a set of criteria for measuring the efficiency of an external audit sample.

Sampling is, however, only one of several testing procedures used by the auditor. The sampling process will be affected by these other forms of test. Therefore, before proceeding to develop a set of criteria for determining the efficiency of audit samples it is first necessary to study the role of sampling within the overall audit process.

2.2 The objectives of external auditing.

In chapter one we noted that the auditor must report to the shareholders on the condition of the company's books of account and on the annual accounts (1). This protects the shareholder against incomplete or inaccurate accounts being presented to him. (Pennington p. 633).

Superficially it might appear that the objectives of the external audit process are clearcut. The auditor is required to ensure that the accounts under audit give a "true and fair view" of the financial state of the company.

However, as we argued in chapter one, the audit objectives are far from clear. A generally accepted model to define and measure audit objectives has not been agreed by the accounting profession.

The auditor is supposed to provide assurance that the accounts under audit give a "true and fair view" of the financial state of the Company but since no generally accepted interpretation of the words "true and fair view" exist, the words are open to a wide variety of interpretations.

In particular there is no generally accepted formula for calculating the value of a "material error" in the accounts, the existence of which would demonstrate that the accounts did not present a "true and fair" view.

This imprecision of objective extends beyond the immediate objectives to the methodology of audit.

Since guidance is seldom given as to the extent of test the auditor is given a wide latitude in this respect.

At one extreme the external auditor could check every single financial transaction under audit to see whether it was valid, accurate and complete and properly posted into the final accounts. Unfortunately the cost of such an approach would be exhorbitant and so the auditor must be selective in his audit. He will not audit every transaction. He will discriminate between transactions and between the various areas under audit. Some transactions and some areas will be checked in great detail, other transactions and whole areas of audit will be lightly checked or no checks will be applied.

This process of discrimination lies at the heart of the audit process, and since discrimination implies sampling, sampling lies at the heart of the audit process.

The skill of the auditor resides in his ability to discriminate as to where to apply his tests and how to properly evaluate the results of his tests.

> "The matters which an auditor must investigate in carrying out his audit of a company's accounts and the depth of his investigation may be prescribed by his contract with the company, or if the contract is silent, by the company's articles." (2)
>
> "If neither the contract nor the articles specify the extent of the auditors duties, his duty is to exercise reasonable skill and care in ascertaining the accuracy of the company's books of account, and the accuracy and completeness of its balance sheet and profit and loss account." (3) (Pennington p. 633)

We conclude that the objectives of the external audit process are subject to a variety of interpretations and the auditor is given a wide latitude in selecting his method of audit.

We will argue later in this study that the major contribution of SS to audit work is that it provides a coherent conceptual framework within which audit objectives can be stated more precisely and audit findings evaluated more objectively.

2.3 The overall framework of the external audit.

Sampling is only one part of the framework of external auditing. The audit process consists of several interrelated parts. The inferences drawn from one part of the audit can affect decisions about the other parts.

Decisions about audit sampling can, for example, be affected by prior information about the quality of internal control and information extracted from the previous years audit files.

In order to find out about the overall framework of external auditing we studied the audit manuals of five multinational audit firms. The introductory section of these manuals provided a basic conceptual foundation to the method of audit.

Each manual described the various stages of the audit.

We identified several important differences in audit method but the basic stages and sequences of approach were similar. The basic stages of the external audit process would seem to follow the sequence:

1. Analytical review.
2. Procedural analysis and testing.
3. Substantive testing.
4. Evaluation of results.

The activities involved at each of these stages are as follows.

2.3.1 Analytical Review.

The auditor or audit team (a) examines the final accounts to detect any unusual items or abnormal pattern of ratios, (b) studies the previous accounts, the previous years audit papers, the internal audit reports and any other information pertinent to assessing the quality and accuracy of the accounts under audit (5).

The assurance derived at this stage will affect the reliability levels applied at the later sampling stage.

2.3.2 Procedural evaluation and testing.

The accounting procedures employed including the accounting controls ought to be documented in a procedures manual.

The auditor will examine this manual with a view to assessing the quality of the procedures and controls built into the accounting system.

Once the auditor has assessed the adequacy of the accounting controls which are purported to be built into the system, he will proceed to test the application of these controls in practice. This will normally be effected by selecting and testing a sample of procedures. (4)

Two types of sampling can be employed. The first type called "sampling in depth" does not usually employ statistical methods. A few accounting documents such as a wage slip, an invoice or a credit note are checked through the entire system of procedures and controls. This approach tests that the procedures manual has been followed on the documents tested but the method does not provide a statistical evaluation of the procedural error rate.

The alternative method is to draw a sample from the population of procedures and use statistical methods to calculate the sample size and evaluate the sampling results.

This latter method is called <u>compliance testing</u> of procedures. A description of SS methods suited to compliance testing is presented in chapter four. A critique of these methods is presented in chapter eight.

2.3.3 <u>Substantive testing</u>.

The final accounts are made up from a set of values. These values are either stocks or flows.

As was explained earlier, the auditor, for reasons of economy, is seldom able to check every transaction underlying these accounting values. The auditor must select a sample from the total population of value transactions and check this sample. He then extrapolates the conclusion from this sample test to the total population.

If the amount or characteristic of error found in the sample is thought to be significant the auditor will require additional tests to be applied to the population under audit.

Under traditional audit sampling procedures the definition of "significance" is left rather vague and is certainly highly subjective.

It is claimed that statistical sampling can provide a more objective and consistent method of extrapolating conclusions from the substantive audit sample to the population from which the sample is drawn.

Several SS techniques have been developed for applying statistical methods to substantive testing of value. A description of these techniques is provided in chapter four and a critique in chapter eight.

2.3.4 Evaluation of audit results.

After steps (1) to (3) are completed the auditor is in a position to review the results of his analysis and tests.

He must decide whether, in his opinion, the accounts under audit provide a true and fair view of the financial state of the company and are in accordance with the requirement of the law.

SS can provide an estimate of the upper limit on error value remaining in the population after audit. By comparing this amount to the material amount of error set for this set of accounts the auditor can arrive at a decision on whether to accept the accounts or require further action to be performed.

The duties of the external auditor are not exclusively concerned with containing error value, but a system which provided an objective upper limit on error value must be considered as making a major contribution to audit practice.

Exhibit 2.1 illustrates the cumulative probability that the total error value in the population exceeds a given amount. In the case illustrated the auditor is 97% sure, in the statistical sense, that the value of error remaining in the population <u>after</u> the audit is completed does not exceed £80,000.

2.4 The role of sampling in auditing.

As stated above sampling is an important technique at two stages in the audit process. These are the stages we defined as procedural testing and substantive testing.

If sampling is to be used, it seems reasonable to test the proposition that scientific sampling using statistical methods can

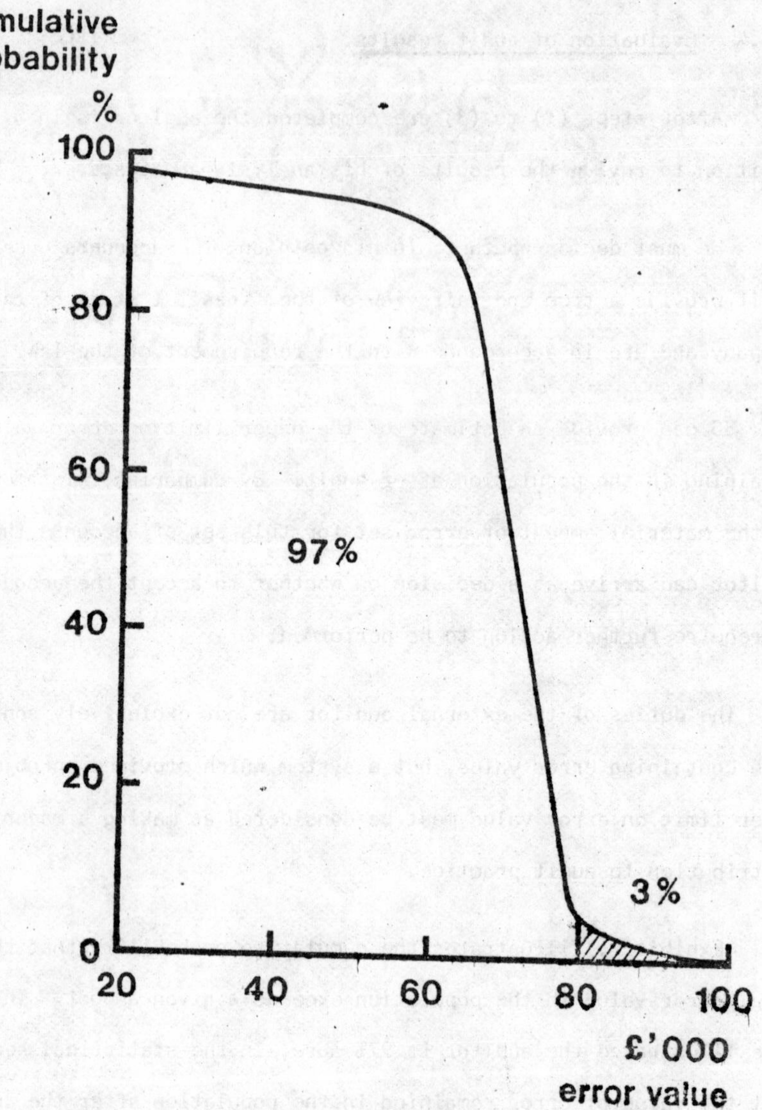

Exhibit 2.1 The diagram illustrates the cumulative probability of the total value of error in the population under audit exceeding £x.

provide a more efficient sample than traditional audit sampling which does not use statistical method.

The main purpose of this study is to examine the proposition that statistical sampling provides, on the average, a more efficient sample than traditional sampling.

We must never forget, however, that sampling is only one part of the audit process which contributes towards the auditors overall assessment of the quality of the account. (See Kinney 1977).

2.5 The objectives of audit sampling.

When statisticians look at the audit process they tend to treat it as a straight forward hypothesis testing operation. The hypothesis to be tested is that the error rate or error value in the population under audit is above some given percentage or amount. A random sample is drawn from the target population and the error found in the sample decides whether the auditor should accept or reject his prior hypothesis about the amount of error in the population under audit.

This statistical approach to audit sampling is not incorrect but it is also not the complete story. The statistical approach assumes that the sole objective of audit sampling is to draw a representative sample from the population under audit. The ideal audit sample would thus be a miniature replica of the population from which the sample is drawn.

Ijiri and Kaplan (1971) have questioned this assumption about the objectives of audit sampling by pointing to the multiple objectives imposed on external audit sampling. Ijiri and Kaplan postulate that an external auditor has four objectives when he designs his audit samples. The four objectives they cite are that the audit sample should be representative, corrective, preventive and protective.

In a later chapter of this dissertation we will assess the efficiency of various statistical sampling methods against these and other criteria so let us, at this stage, examine the Ijiri-Kaplan objectives in detail.

2.5.1 Representative.

The audit sample is representative, in an ideal sense, if it proves to be a perfect miniature replica of the population from which it is drawn. Inferences extrapolated from such a sample to the population from which it is drawn will be correct. A perfect replication of the population is unlikely but a random sample of sufficient size should provide a replica close enough to satisfy the auditors needs.

Stratification of the population prior to sampling may improve the likelihood of a representative sample being drawn.

The more representative the sample the better will be the inferences drawn from the sample. A statistician, as we noted above, is likely to cite this objective as being the sole criteria of an efficient sample. Ijiri and Kaplan, however, suggest that audit sampling is unusual in having the following secondary objectives.

2.5.2 Corrective.

The auditor may wish to modify the design of his total sample in order to increase the chance of identifying and correcting errors.

Such a sample would not be a representative sample since it would be likely to contain a much higher proportion of errors than a representative sample.

An auditor will only carry out this secondary sampling operation if he believes that he can correct a significant number or value of errors. Chapter three suggests that the fraction of the total value of error in the sample is likely to be so small that the total value of correction is not likely to be significant in most audits. But there can be no doubt that in some audits the desired correction can be significant.

It is important to repeat that a sample designed to maximise the discovery of error is not a representative sample and so cannot be used to make inferences about the population from which it is drawn.

2.5.3 Preventive.

Ijiri and Kaplan use the words "preventive sample" to mean a sample drawn in such a way that it is difficult for the auditee to predict the items which will be drawn.

Smurthwaite (1968) and others have demonstrated that traditional audit samples can to some extent be predicted i.e. not sampling the same month two years running, higher probability of sampling busy periods etc.

Statistical sampling requires that the audit sample within each stratum be a random sample and that every item under audit is within the sampling frame. Therefore statistical sampling insists that the audit sample is a preventive sample. It is advocates of traditional sampling who must prove their audit samples to be "preventive".

2.5.4 Protective.

The final objective of audit samples suggested by Ijiri and Kaplan is that the sample be protective. By protective, Ijiri and Kaplan mean that the auditor attempts to include as high a proportion as possible of the total value of the population being audited in the sample. He thus protects himself against the charge of negligence if a significant error is subsequently found in this population.

Conventional statistical sampling methods will not maximise this ratio but traditional sampling methods will tend to do so by sampling all large value items.

In recent years statistical audit sampling methods have been adapted to solve this problem. Two approaches have been used. First the population under audit has been stratified by value and a higher proportionate sample applied to the high value strata. Secondly a form of scientific sampling called probability of selection proportionate to size (PPS) has been applied to audit samples. These methods will be described in detail in a later chapter.

Thus both traditional sampling and statistical sampling can now satisfy the protective condition put forward by Ijiri and Kaplan.

We conclude that, of the four necessary objectives of audit sample design put forward by Ijiri and Kaplan, some form of statistical sampling can be proved to satisfy three objectives namely the representative, preventive and protective objectives. Traditional audit sampling can only, at best, satisfy two objectives namely the corrective and protective objectives.

We should note in particular that a traditional sample which satisfies the corrective objective cannot be a representative sample and so cannot be used to make any inferences about the population from which it is drawn.

We can put this another way by saying that a sample designed to satisfy the corrective objective cannot satisfy the representative objective and is most unlikely to satisfy the protective and preventive objectives.

In our opinion this conflict of objectives ought to persuade the external auditor to ignore the corrective objective when designing his initial sampling plan. If the results of his sample suggest that correction is necessary he must then design a second sample to maximise the probability of identifying errors.

2.6 Some further objectives of audit sampling.

The Ijiri-Kaplan criteria for deciding the objectives of an audit sample give us a set of criteria for measuring the efficiency of the sample. There are, however, other criteria for measuring sampling efficiency which stem from the audit process itself.

We shall call these additional criteria consistency, operational simplicity and economy.

2.6.1 Consistency.

Other things being equal it is important that the standard of investigation in audits should be consistent between different audits and different auditors. Otherwise some audits will be over-investigated and others under-investigated.

Case	Sample Size Range	Median
A. Substantive test - accounts receivable:		
(Total population - 1144 accounts)		
Strong control - positive confirmations	0 to 606	45
- negative confirmations	0 to 1144	15
Weak control - positive confirmations	0 to 1005	101
- negative confirmations	0 to 1144	81
A detailed analysis of this case study is included on pages 10 to 12 of Appendix A.		
B. Substantive test - inventories:		
(Total population - 250 items maximum)		
Test counts	2 to 112	48*
Pricing	9 to 120	42*
Extensions	1 to 128	44*
* Means rather than median. Includes all responses.		
C. Substantive test - prepaid expenses:		
(Total population - 178 balances)	0 to 178	22
D. Compliance test - purchases:		
(Total population - 3692 vouchers)		
Using judgemental (i.e. non-statistical) techniques	10 to 923	45
Using statistical techniques	15 to 343	115

Exhibit 2.2 A test of uniformity in selecting size of audit sample.

Note the wide disparity in size of sample selected under identical conditions.

Source. CICA Study. (1980).

Under traditional audit practice it is difficult to <u>prove</u> a consistent standard of audit. As noted above, the CICA carried out an investigation, CICA (1980), on the extent of audit practised by different auditors in Canada and found a wide variation in standards applied. (4)

The results of a test of uniformity regarding audit sample size is demonstrated in Exhibit 2.2. Four case studies A, B, C and D were presented to 214 audit firms. The firms were asked to state the size of sample they would draw under the given circumstances. As demonstrated in Exhibit 2.2 the lack of conformity in audit sample size, in every case, is surely disturbing?

A statistical approach to audit sampling (and to a lesser extent analytical review) permits the setting of a consistent standard of audit. Once the prior requirements of the auditor are known a statistical approach allows a consistent standard to be applied to the audit sampling process. This consistent standard ensures that a given total audit sample will be applied in an optimal fashion over several accounting populations.

However, as Case D shows, agreement on initial requirements as to confidence and precision are needed before even SS will provide a consistent sample size.

2.6.2 Operational simplicity

If a sampling method is complicated to operate this will count against it, unless a computer and suitable computer audit package are available.

Even when a computer is available a complicated sampling scheme may raise problems in evaluating the sample result.

Complicated sampling schemes tend to raise tricky conceptual problems when evaluating the estimator.

If a sample can be selected and evaluated manually this must count in its favour.

Note that a sophisticated sampling plan can be applied in a simple way.

2.6.3 Economic efficiency.

When an auditor draws and checks a sample he will derive a certain amount of information from this sample.

Under traditional auditing it is difficult to relate sample size to specific audit objectives. Even if the auditor defines his objectives with great precision he is unable to relate these objectives to a specific sample size. Why does he draw a specific sample size of n units from the population under audit rather than any other number of units?

Statistical sampling allows the auditor to relate the sample size to the specific requirements of the audit test. Once materiality and required confidence in the inference are stated, the auditor, by using statistical method, can calculate the minimum audit sample size required to meet those specific requirements.

Since sample size is minimised, sampling cost is likely also to be minimised (6). Much ingenuity has been shown by statisticians in devising sampling methods which achieve given audit requirements at minimal sample size.

Only a statistically derived audit sample can be <u>proved</u> to be the minimum sample size needed to satisfy the given sampling objectives.

However this leaves one economic problem unresolved.

Suppose the minimum sample size required to meet the auditors prior stated requirements is too large? That is the cost of sampling exceeds the part of the audit fee allocated for audit sampling. What is the auditors next step? In theory he should appeal to the shareholders for an additional fee. In practise this is an improbable course of action. We will return to this important caveat in chapter ten.

For the moment let us define a strong and weak criteria for measuring economic sampling efficiency in auditing.

The <u>strong criteria</u> is defined as being the minimum sample size needed to meet the specific audit objectives. These are determined by the required reliability and confidence interval.

The <u>weak criteria</u> is defined as being a sample size small enough to fall within the range of sample size traditional in auditing.

Statistical audit samples can be shown to satisfy the strong criteria but possibly, not the weak.

Traditional audit samples will, by definition, satisfy the weak criteria but not the strong.

2.7 <u>Alpha and beta risk.</u>

Once the audit sample has been selected and evaluated the auditor must decide whether to accept or reject the accounting population under audit based on his evaluation of the sample results.

Whichever decision he makes he takes a risk. Statisticians call this risk, alpha risk and beta risk.

There is much discussion of alpha and beta risk later in this study.

Alpha risk is defined as the risk that the auditor will reject a population when, in fact, he should have accepted it.

Beta risk is defined as the risk that an auditor will accept a population when, in fact, he should have rejected it.

The relative importance of alpha and beta risk depends upon the opportunity cost of getting the decision wrong.

There can be no doubt that in the context of external auditing beta risk is more important than alpha risk.

Beta risk incorporates the possibility of the auditor being sued for negligence.

Alpha risk incorporates only the cost of overauditing (by the auditor or his client).

Several methods of SS appear to ignore alpha risk. (Elliott and Rogers 1972 and Kaplan 1975 discuss the problem). We do not recommend ignoring alpha risk. The problem of assessing alpha and beta risk will be discussed in more detail in chapter eight.

2.8 Non-sampling risk.

Alpha risk and beta risk are concerned with sampling risk. The auditor also faces a degree of non-sampling risk when he deduces an inference from an audit sample.

Non-sampling risk occurs because of the ignorance or the negligence of the sampler. He samples in the wrong way or draws invalid conclusions from the sample. (See Roberts 1979).

Sampling by computer greatly reduces the probability of non-sampling risk but it can never be entirely discounted. A control should always be applied to audit the accuracy of the sampling process.

Clearly, complicated sampling systems increase the non-sampling risk.

2.9 Conclusions.

In this chapter we have searched for a set of criteria for measuring the efficiency of an external audit sample.

Efficiency can only be measured in relation to specific objectives and so to define a set of criteria for measuring the efficiency of audit samples we must first define a set of objectives for the external audit process.

We did not find this easy to do. A fully developed conceptual framework of the external audit process, which is generally accepted, is not available.

We also discovered that there is no generally accepted standard procedures in the methodology of audit. There does, however, appear to be general agreement on the four basic components of the audit process, namely analytical review, evaluation and testing of procedures, substantive testing and evaluation.

We noted that it is not economically feasible to audit every transaction underlying the final accounts and so an important part of an auditors skill lies in his ability to discriminate between areas and items requiring investigation. Discrimination implies sampling and so sampling lies at the heart of the audit process.

Statistical method can be applied at all four stages of the audit process and statistical sampling at stages two, three and four.

The key question is whether a statistical sample can be demonstrated to be more efficient than a traditional audit sample.

In order to answer this question we must devise a set of criteria for measuring the efficiency of an external audit sample.

We chose the set of objectives put forward by Ijiri and Kaplan (1971) as our initial guide. IK suggest that an external audit sample ought to satisfy four objectives, the sample is required to be representative, corrective, preventive and protective. We note a conflict of objective between the corrective criteria and the others, which suggest a two stage sampling process.

In addition to the Ijiri Kaplan objectives we suggest a further three objectives stemming from the audit process.

These additional objectives are consistency, operational simplicity and economic efficiency.

The last of these we further subdivide into the two categories of strong and weak efficiency.

We conclude that a statistical audit sample can be proved, on the average, to meet all these objectives, except the corrective one. It is difficult to prove that a traditional audit sample is representative, preventative, consistent or economically efficient.

When designing his sampling plan the auditor is concerned both with alpha risk and beta risk. In an audit context we believe beta risk to be the more important.

The auditor should also concern himself with minimising non-sampling risk. Sampling by computer reduces this particular form of risk.

In this chapter we have suggested a set of objectives to bear in mind when designing an audit sample.

In later chapters we will use these objectives as a set of criteria for measuring the relative efficiency of various audit sampling methods.

Notes on Chapter Two.

1. Companies Act 1967 s 14 (1).

2. Re City Equitable Fire Insurance Co. Ltd. (1925) Ct. 407 at 499.

3. Re London and General Bank (No. 21) 1895 2.

4. CICA "Extent of Audit Testing" (1980 pp. 18-19).

5. The audit team normally consists of a partner, manager, senior auditor and several junior auditors. The team will usually have access to further specialists, if required.

6. Making the reasonable assumption that an auditor puts equal effort into auditing each unit sampled. If he puts in less effort per unit for a larger sample, sample size is not related in total testing cost. This problem has not been followed up in the literature. There may also be a trade off between set-up time for a complicated system and testing time on a smaller sample.

7. However it should be noted that when an auditor is simply seeking assurance that the error rate is less than some relatively high upper bound, say 5%, then the sample size increases approximately linearly to the tightening of the acceptable error rate. (We thank the reviewer for this note).

Chapter 3.

ERROR PATTERNS IN ACCOUNTING POPULATIONS.

3.1 A suggested scheme for classifying errors found in accounting populations.

Error is endemic to accounting populations. The objective of auditing is to ensure that the degree of error in accounting populations is kept within acceptable bounds.

In the context of accounting the classification of error is not simple. Accounting error has several attributes and at least two dimensions.

Taylor (1974) provides a useful classification of the various types of error encountered in accounting populations.

Exhibit 3.1 suggests a simple scheme for classifying any accounting error into one of two categories. The error is either an error of principle or an operational error, the error is either accidental or deliberate.

This classification scheme allocates all accounting errors to one or other of the four boxes set out in exhibit 3.1.

This study is primarily concerned with errors falling into box AD, that is it is concerned with operational errors which are committed accidentally.

Fraudulent error is likely to concern the auditor in the context of professional liability for negligence. Accounting errors committed deliberately, that is mainly fraudulent errors, are specifically excluded from this study. The pattern and incidence of fraudulent errors are likely to be quite different from those of accidental error.

	Error of Principle (C)	Operational Error (D)
Accidental (A)		
Deliberate (B)		

Exhibit 3.1 A scheme for classifying accounting errors.

Statistical methods can assist in detecting fraudulent error but the techniques of SS suited to detecting deliberate error will be very different from the techniques suited to estimating an upper bound on accidental error (negligence).

Sampling is unlikely to be used for detecting errors of principle. Errors of principle are usually detected during the analytical review stage of the audit.

Statistical sampling methods are best suited to assuring the auditor that the incidence and value of error remaining in the accounts after the audit is completed is kept within acceptable bounds.

3.2 Other aspects of accounting error.

There are, however, certain other aspects of operational error which do concern us.

In particular we must differentiate between compliance error and monetary error.

A compliance error occurs when an accounting rule has been breached. A compliance error need not involve an error of value.

A monetary error occurs when the wrong value is assigned to an accounting number.

Most substantive errors are monetary errors but a compliance error could be substantive.

Secondly we must differentiate between a biased and unbiased type of monetary error. A biased error is sometimes called a systematic error.

If an error is unbiased, and sufficient of these unbiased errors occur, they will tend to cancel themselves out: The mean

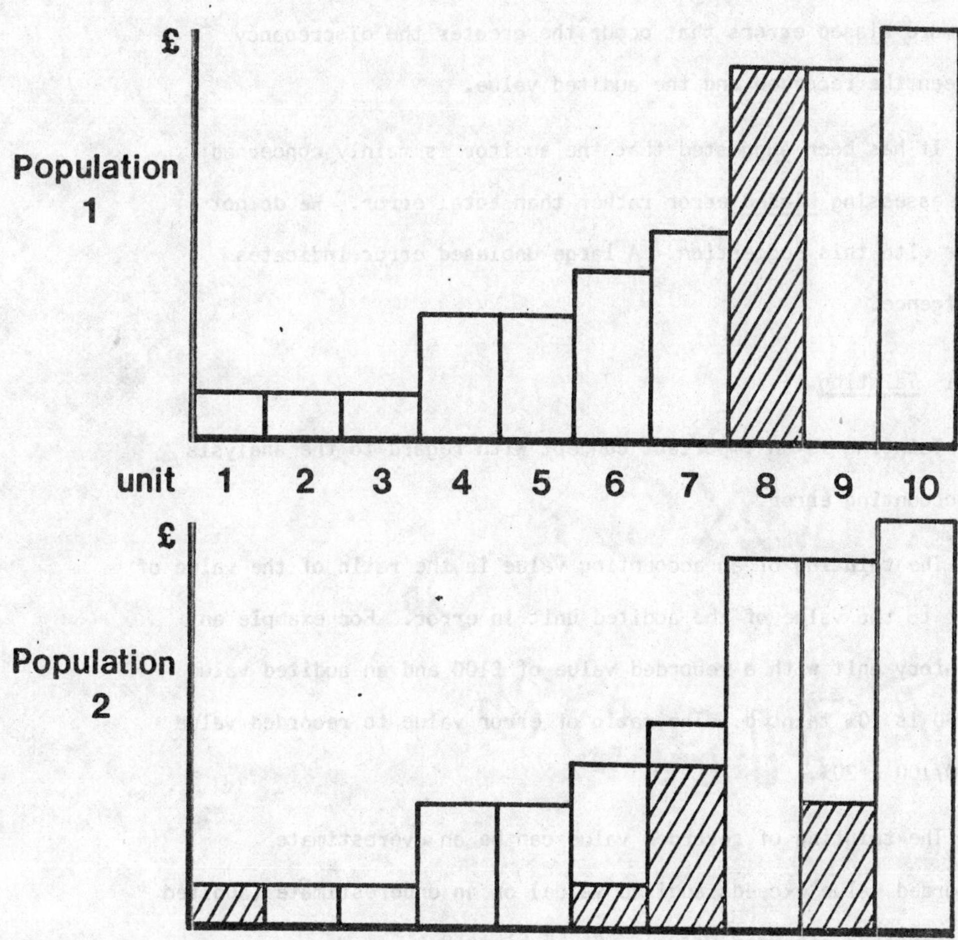

xhibit 3.2. **ERROR TAINTING.** The diagram illustrates the different tainting distributions of a sample error value of £9 in two audit samples.

= overstatement error.

value of a population of unbiased errors will tend towards zero (1). The accuracy of the recorded total value of the population under audit would be little affected by unbiased errors.

On the other hand a biased error always tends to lean one way, the more biased errors that occur the greater the discrepancy between the recorded and the audited value.

It has been suggested that the auditor is mainly concerned with assessing biased error rather than total error. We do not agree with this suggestion. A large unbiased error indicates negligence.

3.2.1 Tainting.

Tainting is an important concept with regard to the analysis of accounting error.

The tainting of an accounting value is the ratio of the value of error to the value of the audited unit in error. For example an inventory unit with a recorded value of £100 and an audited value of £80 is 20% tainted. The ratio of error value to recorded value is 20/100 = 20%.

The tainting of recorded value can be an overestimate (recorded value exceeds audited value) or an underestimate (audited value exceeds recorded value). See Exhibit 3.4.

Exhibit 3.2 illustrates a population of the audited units with different tainting distributions. Both audit samples are tainted by £9. In population 1 the tainting is limited to a single unit, unit 8, which is 100% tainted. In population 2 the tainting is distributed as follows:

Unit	tainting %	Error Amount £
1	100	1
6	25	1
7	80	4
9	33.3	3
		9

3.3 Some statistical attributes of accidental operational error.

As argued above the auditor is concerned with assessing the degree of procedural and monetary error in the account he is auditing.

If he is to use statistical methods to assist in this assessment it would be helpful if he knew something about the likely statistical characteristics of the accounting errors he is trying to evaluate.

What, for example, is the mean error rate of compliance errors? Is the distribution of error values skewed?

The answer to such questions will assist the auditor in selecting a sampling scheme best suited to picking up these patterns of error at minimum cost.

A prior knowledge of the usual statistical attributes of operational error will provide the following benefits:

(a) If an auditor is searching for error patterns which vary from the usual error pattern it is surely necessary for him to know what the usual error patterns are in this kind of account and this kind of client.

(b) Many sampling techniques require that certain assumptions be made about the distribution of errors and error values in the population under audit. If the most likely distribution is notknown then the conservative principle suggests that the worst possible distribution (2) should be assumed. This conservative approach may well increase audit sample size above what it need be if the most likely distribution of error were known. (6)

In other cases the sampling design is crucially affected by prior assumptions about such things as the proportion of understatement errors of value, the relation between error value and the value of the item containing the error, the allocation of total errors between units of high and low value .... and so on.

If the auditor knows the usual pattern or error in such cases he is able to accept a less conservative sample design. This should reduce the cost of sampling.

(c) The Bayesian approach to sampling theory can be employed in auditing if prior probabilities of error rates can be estimated from studies of actual error rates in the type of population under audit. Accurate prior probabilities, using the Bayesian approach, can provide significant savings in audit sample size.

For all of these reasons it is important for an auditor to study the statistical characteristics of errors discovered in accounting populations.

3.4 Definition of some variables used in describing accounting error.

The population of book values being audited will be called the recorded values and will be coded $y_1, y_2 \ldots y_n$. Also $y_1 + y_2 \times y_3 \ldots + y_n = \sum_{i=1}^{i=n} yi = £Y$. Thus £Y is the sum total of the recorded population.

The population of audited values will be coded as $x_1, x_2, x_3 \ldots x_n$ and $x_1 + x_2 + x_3 \ldots + x_n = \sum_{i=1}^{i=n} x_i = £X$. Thus £X is the sum total of the audited population.

An error e_i occurs if $y_i \neq x_i$. The total of differences $e_1 + e_2 + e_3 \ldots + e_n = \sum_{i=1}^{i=n} e_i = £E$. Thus £E is the sum total of momentary error in the population under audit.

Let £M represent materiality. Under normal audit conditions the auditor need not estimate £E but rather attempt the less onerous task of ensuring that £E<£M.

3.5 Knowledge required on the nature of errors in accounting populations.

It was suggested above that the design of audit sampling plans could be much improved if more were known about the nature of error in accounting populations (3).

It would, for example, be helpful if auditors knew the answers to the following questions.

(a) What is the mean rate and variance of compliance error in various types of accounting populations? In a given population is this mean rate constant from year to year or does it vary a great deal?

(b) Does the mean rate of compliance error vary between different kinds of accounting populations and different kinds of clients?

(c) What is the mean rate and variance of error value? Does it vary from year to year? Does it vary between different types of accounting populations and different clients?

(d) Are error values usually biased, usually skewed and do large understatements (4) of error occur often or rarely? Do large understatements normally occur in small value items?

(e) How frequently are zero values or negative values found in various types of accounting populations?

(f) Are error values usually proportionate in size to the value of the item in which they occur?

(g) Is there a higher probability of error in low value units compared to high value units (Because the control system is less rigorous on low value units)?

(h) Is the distribution of total error value within an accounting population different for different types of population?

(i) Are errors distributed randomly in accounting populations or do errors tend to cluster?

Later in this chapter we will suggest tentative answers to some of these questions based on the data displayed in the following section.

3.6 Some empirical data on error patterns found in accounting populations.

Since the number of financial audits performed in any one year throughout the world must run to several million, and since audit activity is over 100 years old, it is odd that so few studies of accounting error have been published.

The first published studies of errors discovered in accounting populations by external auditors would seem to be Johnson, Leitch, Neter (1979) and Ramage Krieger Spero (1979). An early study of errors detected in invoices received was published by Gregory (1951). Otherwise we are dependent on a few theoretical papers which use actual error rates as the basis for their analysis. Cyert, for example, published several papers and Neter and Loebbecke (1975) based their simulation study on actual populations.

Since many sampling techniques are highly sensitive to the rate and distribution of accounting error in the target population more research on this topic is to be recommended.

3.6.1 Early studies providing data on the pattern of accounting error.

Jones (1947) suggested that error rates below 0.3% were "acceptable" and below 0.9% "fair" in clerical work.

Vance (1950) took 0.5% as being acceptable and 3% as being unacceptable error rates in clerical work (compliance error).

Other actual error rates discovered in clerical work were:

Study	Year	Error rate %	Type of Population
Vance	1950	0.5	Inventory
Cyert and Johnson (5)	1957	0.2	Statements of account
Buchan and Cyert (5)	1957	0.25	Invoices
Johnson and Rowles (5)	1957	0.54-1.03	Debtors accounts
Davis and Rounseville	1959	0.25	Freight bills
Burstein	1967	2.00 (of value)	Inventory
Aly and Duboff	1971	2.00 (of value)	Debtors
Neter and Loebbecke	1975	28.6	(Freight) Debtors (low value)
		71.0	Inventory (medium value)
		7.3	Debtors (low value)
		5.7	Debtors (low value)

The Neter Loebbecke (1975) error values were actual but they were not claimed to be typical.

3.6.2 The Gregory study.

R.H. Gregory (1951) studied the pattern of error in invoices received by an automobile manufacturer. Around 35,000 invoices were received during a given period and the error rate was 0.0017% or 17 errors on the average per 10,000 invoices.

The distribution of value of the invoices was highly skew, i.e.

% of invoices	% of value
48	1.4
39	12.1
13	86.5
100	100.0

The 20% of invoices with a value in excess of $500 accounted for 78% of net dollar amount of corrections. Gregory found a "smaller error rate for lower value invoices". Also the error rate was not stable in total or as to value distribution between between the two years studied.

The adjustments required were not symmetrical, credit adjustments (overstatements) were 92% of total value adjustments.

Gregory carried out a similar study on transport invoices. He found that 16% of invoices represented 87% of dollar value. 72% of the number of invoices accounted for 58% of the adjustment value. Most errors were for small amounts.

3.6.3 The PMM report on audit errors.

PMM, a leading firm of professional accountants in the USA, provided a list of the errors discovered in 111 audit files. These errors have been studied by two teams of academics. The preliminary results of this research are published in Ramage, Krieger, Spero (1979) and Johnson, Leitch, Neter (1979).

There appears to be some doubt as to how representative these audit files are. RKS state that "the data should not be regarded as representative of all auditing populations" and "Error rates may well be higher .... than is usual". (p4)

However JLN state that "The CPA firm informed us that the selected audits are intended to be representative of audits for the firm's <u>larger</u> clients for which statistical sampling is utilised". (p5)

The date suffers from other limitations. For example most inventory error rates come from a single industry, several audit populations come from the same client, the researchers had no control over the selection or presentation of the data and the strata were not equally weighted.

This data is, however, one of the most extensive reports on accounting errors discovered by auditors to be made publically available and so the findings of the two studies are highly relevant to our examination of the pattern of errors in accounting populations.

The RKS findings are as follows:

(a) Error <u>rates</u> vary by both type of population and client. The variation is substantial. See exhibit (3.4).

(b) Error rates are not affected by value of units processed. A slight tendency is found for higher error rates to occur in high value units.

(c) The proportion of overstatement errors ($X_1 < Y_1$) and understatement errors ($X_1 > Y_1$) differ for different types of population. For example debtors have few understatements but inventory has a substantial fraction of understatements. See exhibit (3.4(a) and (b)).

The fraction of understatements or overstatement does not seem to be related to either error rate or recorded value. But the value of understatements are nearly always much less than overstatements in the same population.

(d) The absolute magnitude of error typically increased with the value of the unit in error (recorded or audited value). "The general pattern is thus nearly constant error relative magnitude as audit value increased in magnitude". (p21)

The JLN analysis of this same data does not agree with this last conclusion but otherwise follows the RKS findings. In addition the JLN study notes that:

(a) The value of errors in populations of debtors tend to be larger and less variable than the value of inventory errors.

(b) The distribution of error values are far from being normally distributed for either type of accounting population. Accounting error distributions are more peaked and have fatter tails upwards than do normal distributions. They tend to be highly skewed, usually to the right.

One most interesting and promising finding is that "standardised distributions of error values for each type of audited population tend to be highly similar". (p34)

(c) The distribution of error taintings are highly variable. The tainting distribution for inventory was very different from that for debtors. The median tainting for debtors was 48% for inventory - 8%. Inventory errors showed large and quite frequent negative taintings but 100% overstatement errors were rare. debtor error taintings showed the reverse characteristics.

The distribution of error taintings is not normally distributed. Some inventory taintings were negatively skewed.

(d) Error amounts contained in large value units were more variable, particularly for inventory.

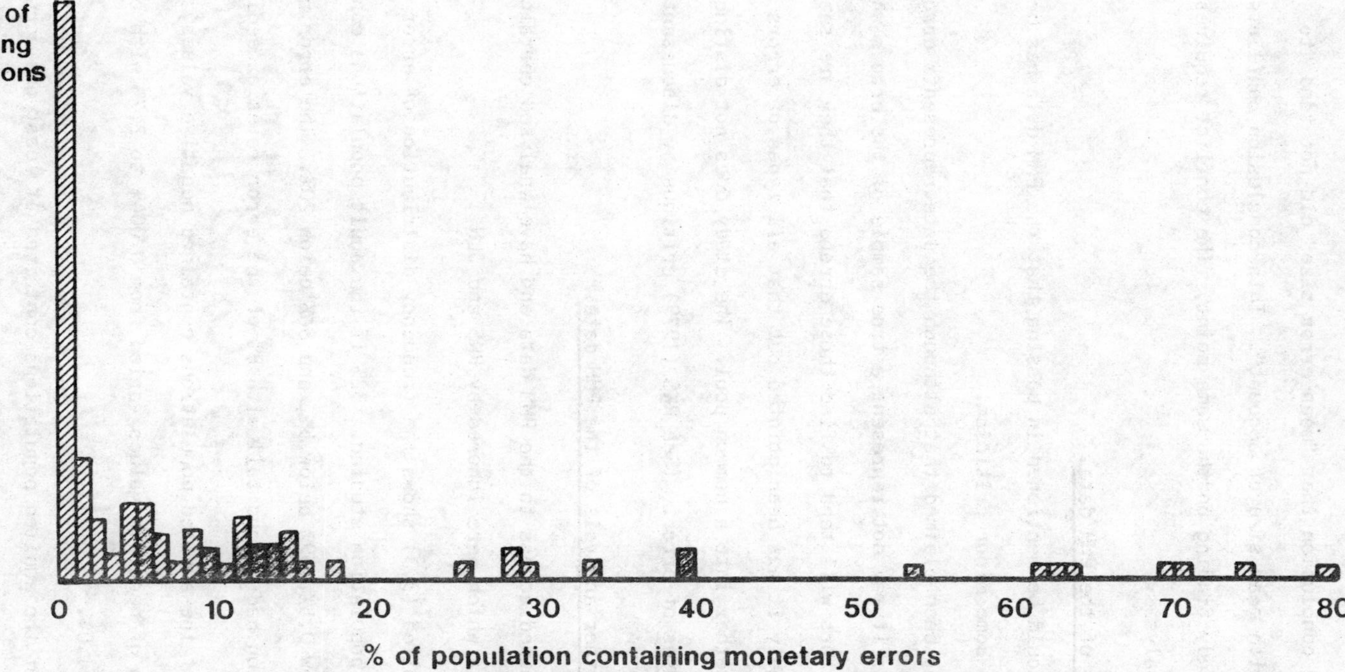

Exhibit 3·3, FREQUENCY DISTRIBUTION OF % OF MONETARY ERRORS DISCOVERED IN 97 AUDIT SAMPLES

Source: PMM

As we noted above, the one disturbing finding of the analysis is the JLN conclusion that "mean error size" did not tend "to increase with mean size of account". This conclusion conflicts with the RKS study finding on the same point. The conflict requires further analysis.

A critique of the PMM data.

It should be mentioned in passing that the PMM data has been subject to some minor criticism.

It has been claimed that although the date represents errors found by auditors it may not represent a true sample of the errors which exist since auditors will tend to find those errors that they are searching for!

Secondly it has been pointed out that all types of errors have been lumped together into a common pool. The study does not distinguish between differ causes of error. (See RKS (1979) critique by discussants).

3.6.4 Further analysis of the PMM data.

We gained access to the PMM data and have analysed certain aspects of the data which were ignored by RKS and JLN.

Exhibited (3.3) shows the frequency distribution of error rates in 97 of the population studied. 32% if the audit populations contained error rates below 0.5%, 59% below 5%, and 86% below 20%. The error rate distribution had a long tail with 8% of the error rates exceeding 50%!

13% of the audited populations contained negative values, the proportion of negative values varied from 0.004% to 2.7% with a median of 1.0%.

51% of the audited populations contained at least one account with 100% error. These were mostly small accounts.

(a) Statistical characteristics of monetary error rates

	Inventory	Debtors	Test of % Significance
Mean	5.2	3.17	99
Median	1.9	2.2	95
Standard Deviation	6.15	2.87	99
Coefficient of variation	1.18	0.91	95
Skewness	+ 1.61	+ 1.01	99

(b) Overstatement and understatement errors.

	Inventory		Debtors	
	No.	%	No.	%
O/s	43	80	94	89
U/s	11	20	12	11
Total	54	100	106	100

Difference significant at 99% level.

Exhibit (3.4) Some findings from the CG study of accounting errors found in audited populations. All errors were monetary errors.

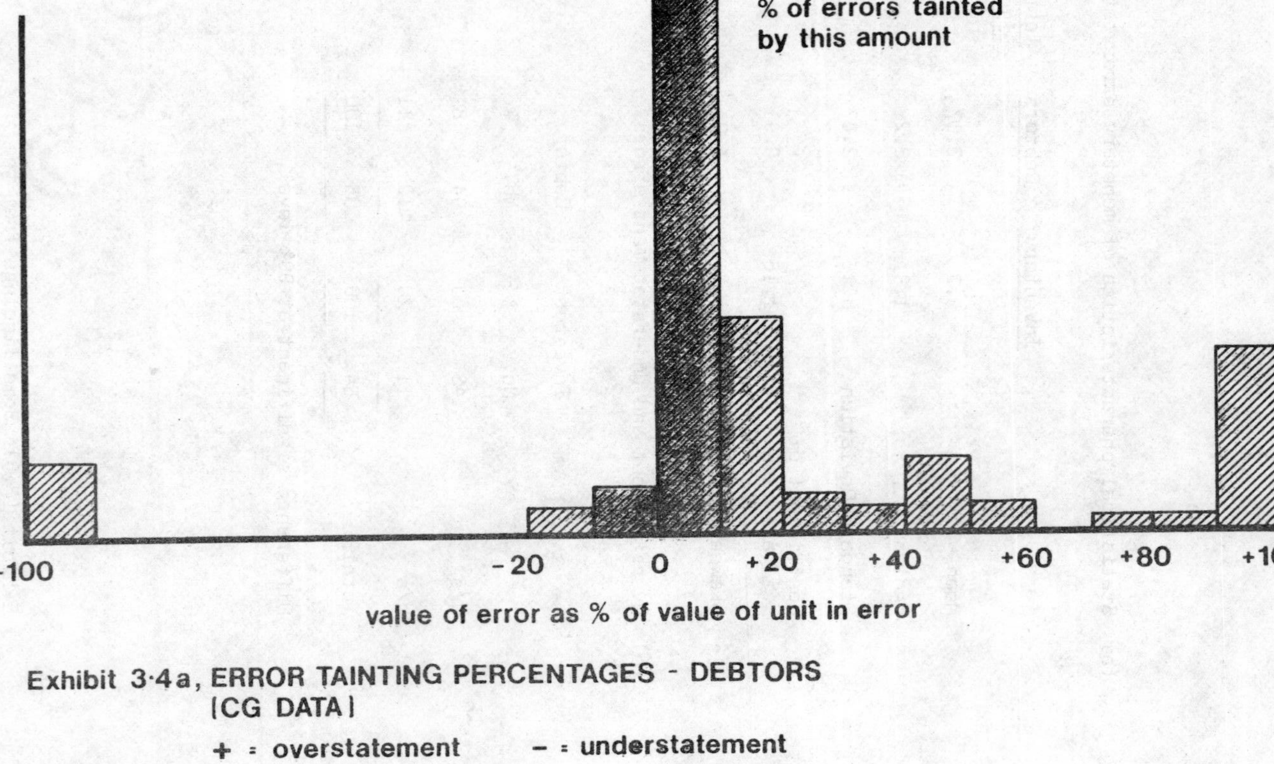

Exhibit 3·4a, ERROR TAINTING PERCENTAGES - DEBTORS
(CG DATA)

+ = overstatement — = understatement

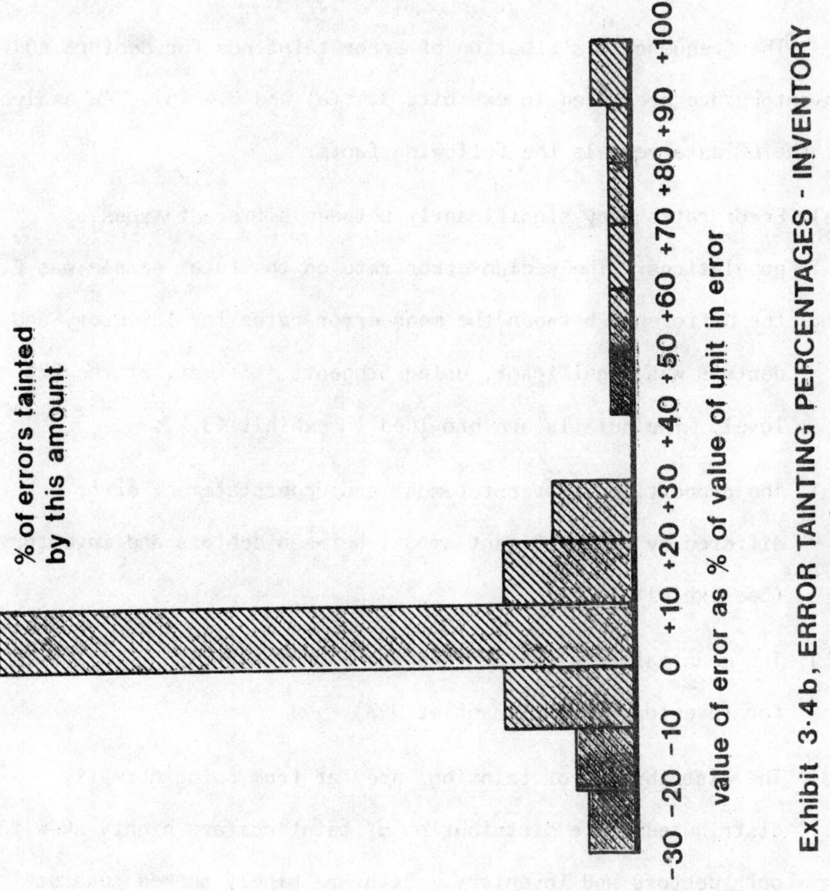

Exhibit 3·4 b, ERROR TAINTING PERCENTAGES - INVENTORY (CG DATA)

+ = overstatement − = understatement

3.6.5 The CG report on accounting errors.

CG, a leading firm of professional accountants in Canada, provided an analysis of taintings of monetary errors discovered in audited accounting populations (Leslie 1977).

This list was not as detailed as the PMM listing but the similarity between the pattern of error taintings is striking.

The frequency distribution of error taintings for debtors and inventory are presented in exhibits 3.4 (a) and 3.4 (b). An analysis of the CG data reveals the following facts:

(a) Error rates vary significantly between different types of populations. The median error rate on the total sample was 2.1%. the difference between the mean error rates for inventory and debtors was significant, using Students 't' test, at the 99% level. The details are provided in exhibit (3.4).

(b) The proportion of overstatement and understatement error differed by a significant amount between debtors and inventory. (See exhibit (3.4).

(c) The mean tainting for debtors was higher than the mean tainting for inventory (significant at 99%).

(d) The distribution of taintings are far from being normally distributed. The distribution of taintings are highly skew for both debtors and inventory. Both are highly skewed towards overstatements.

(e) The distribution of error taintings:

(1) are highly variable but above 50% of taintings are less than 20%.

(2) are different for different types of population.

(3) demonstrate that inventory has a significant number (20%)

of small understatement taintings. Debtors have fewer understatement taintings but some of those identified were large.

These findings correlate quite closely with the findings in the PMM study.

3.6.6 A United Kingdom Study.

Since both the previous studies were based on audit samples drawn in North America it was decided to attempt a similar pilot study based on audit samples drawn in the United Kingdom.

Method of collecting data

Two large accounting firms in the UK were asked to draw a random sample of twenty audit files each from audits having a year ending in the period 1979-1980. The audit papers on inventory and debtors (bills receivable) were extracted from these files and all errors noted in the audit papers were listed.

The particular characteristics of each error noted were as follows.

1. Compliance or substantive error.
2. Value of item in error.
3. Value of error.
4. Cause of error.
5. Error and understatement or overstatement of book value.

In addition the following information was listed (if available). Industry, type of accounting population, number of items in, and value of, population, sample size and value. Altogether 323 errors were listed from 76 accounting populations - 44 inventory, 32 debtors. 15 audit samples contained no errors. 61 audit samples contained one or more errors. The size and value of sample was not provided in all cases. Details of the audited populations are listed in Exhibit 3.5.

a)	Number of accounts	42
b)	Number of populations audited	76
c)	Made up of:	
	inventory	44
	debtors (bills receivable)	32
d)	Total value audited	£170m
e)	Total value of errors discovered in the audit samples drawn	£765,813
f)	Ratio of (e) to (d)	0.45%
g)	Modal sample size	~ 40

Exhibit 3.5. Details of populations audited

Number of errors		323
Made up of:		
compliance errors	25	
substantive errors	298	323
Errors of:		
overstatement	176	
understatement	122	298

Exhibit 3.6 Classification of number of errors discovered in audit samples.

Note: 40 of the 298 substantive errors were not certified as to value of error.

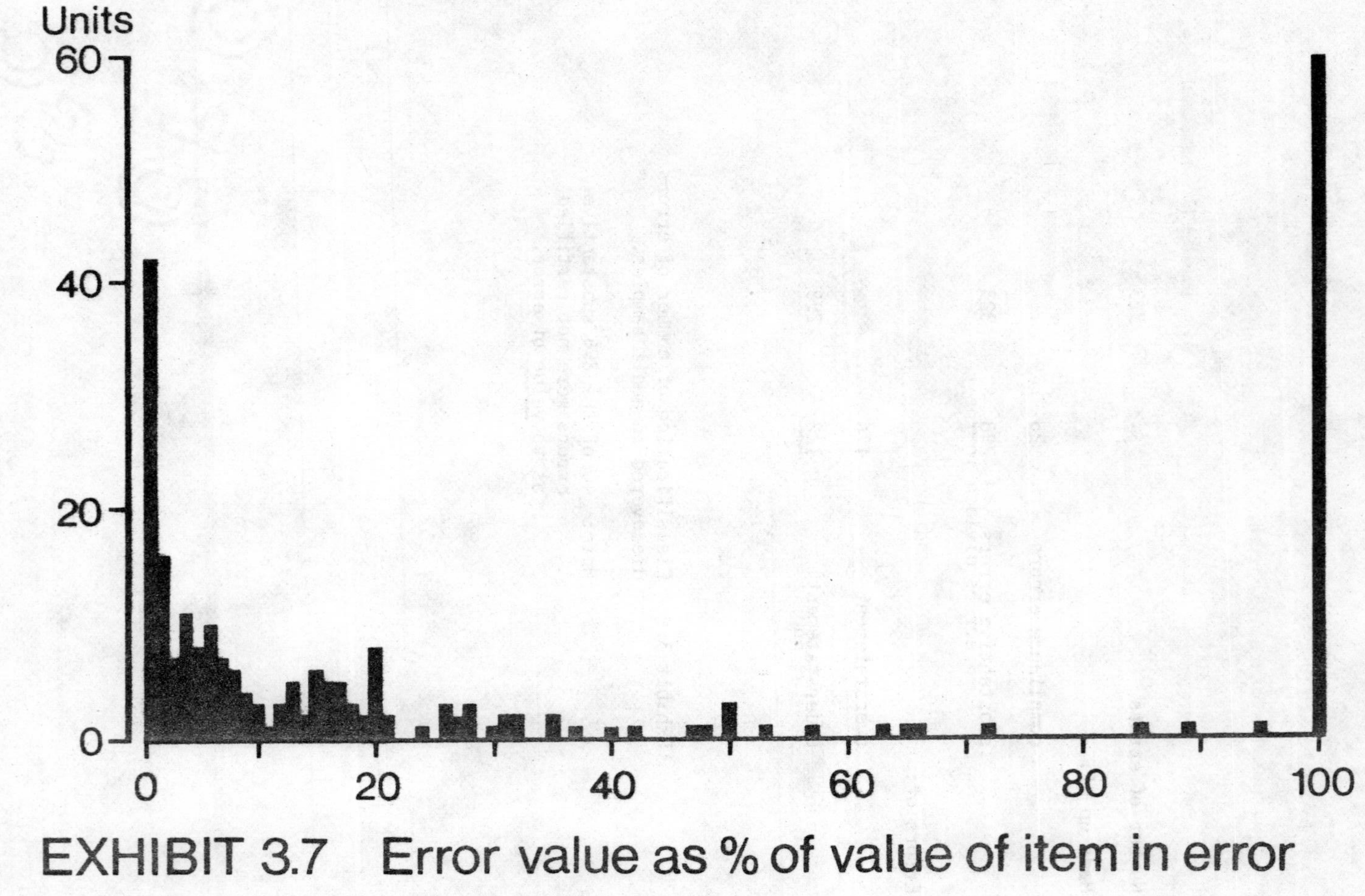

EXHIBIT 3.7 Error value as % of value of item in error

The results

The results of our study of the pattern of error discovered in 76 audit samples is as follows.

Number of errors

As noted in Exhibit 3.6 of the 323 errors noted only 25, a mere 8% were compliance errors, with no monetary affect.

The substantive errors divided in the ratio 6 : 4 between overstatement of book value and understatement of book value.

Tainting of items in error

The term "tainting" is used in audit sampling to describe the ratio between the value of error and the value of the item in error. An item of £60 containing a £15 error is said to be 25% tainted.

A frequency distribution of the tainting percentages of 258 substantive errors (7) is presented in Exhibit 3.7. We note that the distribution is U shaped. The vast majority of taintings being either 100% or below 10%.

Distribution of error values

Exhibit 3.8 exhibits the frequency distribution of the values of errors discovered. The total value of all errors discovered is £765,813. We note that two errors totalling £320,000 make up 42% of the total error value. The 15 errors in excess of £10,000 each make up 83% of the total value of errors discovered. That is 5% of the errors account for 83% of the value of errors discovered. The distribution of error values is highly skew. This finding conforms with that of the other studies mentioned earlier.

Value of error £	No. of errors	% of total value
1 - 100	77	-
101 - 1,000	112	7
1,001 - 10,000	54	10
10,001 - 100,000	13	41
100,000 - 200,000	2	42
	258	100

Exhibit 3.8 Frequency distribution of error values

Errors of	Number of errors %	% of total value of error
Overstatement	59	54
Understatement	41	46
	100	100

Exhibit 3.9 Errors classified between those which overstate book value and those which understate book value

Tainting %	Audited items Exceeding £10,000 %	Less than £2,000 %
0 - 1	35	3
>1 - 10	33	17
>10 - 20	5	19
>20 - 99	17	21
100	10	37
> 100	0	3
	100	100

Exhibit 3.10 Tainting percentages: a classification by relative size the item in error.

The probability of finding a given tainted % appears to be affected by the relative size of the item in error.

Exhibit 3.9 examines the direction of the error value. As noted above the errors divide 6 : 4 in favour of overstating book value, but the total value of error divides more evenly between overstatement and understatement.

Are tainting percentages significantly affected by the relative size of the item in error?

We classified the tainting % into two groups. Those discovered in audited items exceeding £10,000 and those discovered in audited items of less than £2,000.

The results are shown in Exhibit 3.10. Small value audited items had a much higher share of the large taintings, particularly the 100% taintings.

Analysis of the cause of errors.

Exhibit 3.11 attempts to classify the population of errors by cause. Why was the item audited considered to be in error?

Most of the descriptions of error in Exhibit 3.11 are self-explanatory. "Incorrect recorded value" means that the recorded value did not agree with some independent valuation such as a stock count or debtors confirmation. "Cut off date error" means that the wrong cut off date was used at year end.

We noted above that two errors accounted for 42% of the total value of error. One of these was an error of principle. The other an incorrect transcription from one document to another. This latter is listed under incorrect recorded value.

We note the numerous causes of error, the fact that most of these errors are not individually of significant value, and the difference in frequency of the various causes of error between two types of population investigated.

Commentary

First there must be a caveat that this study of error patterns discovered by external auditors in audited accounting populations is based on a small sample from only two firms. The sample is too small and too narrow for the findings to be extrapolated as representative of all error patterns in all accounting populations.

The results of many similar studies must be combined before it can be claimed that a standard error pattern has been identified at an acceptable degree of statistical significance.

It is hoped that this study may encourage further work on the topic in other accounting firms, on other accounting populations, and in other countries.

Some notable findings from the present study are as follows.

Correction

The absolute value of the corrections, at less than one half of one percent of the value of the populations being audited, is hardly likely to be "material" in the audit sense of material.

When we net off the under against the overstatement of book value, the net amount is insignificant.

Therefore, with regard to the entire field audited, the purpose of the audit would seem to be extrapolation rather than correction.

In only 3 out of 76 populations audited did the absolute value of corrections exceed 2% of the value of the population audited.

If the primary purpose of external auditing is not correction of errors but extrapolation of the value of error discovered from the audit sample to the population being audited then a formal

Cause of error	% of total value	Number of errors		
		Total	Inventory	Debtors
Item incorrectly priced	6.7	59	59	0
Incorrect recorded value	28.9	39	31	8
Summation error	2.1	31	27	4
Costing calculation incorrect	6.1	28	28	0
Cut off date error	13.7	26	15	11
Posting to wrong account	9.5	14	6	8
Credit note incorrectly treated	0.8	13	0	13
Dispute over transaction	2.3	10	0	10
Item omitted from listing	2.6	9	7	2
Wrong invoice copied from	0.1	7	3	4
Wrong valuation principle used	24.5	6	5	1
Invalid write off	1.6	3	3	0
Overpayment	*	3	0	3
Incorrect extension	*	2	2	0
Double entry	0.2	2	1	1
Incorrect % used	0.3	2	2	0
VAT error	*	2	0	2
Discount % incorrect	*	1	0	1
Wrong currency conversion rate used	0.6	1	1	0
Totals	100.0	258	190	68

* less than 0.1%

Exhibit 3.11 Analysis of errors by cause

process of extrapolation ought to be used.

Empirical studies of audit practise in the USA, Canada and England and Wales (8) have found that in the majority of external audits the extrapolation process is intuitive.

Variety of error

The study demonstrates the wide variety of types of error inherent in accounting populations (Exhibit 3.11).

When statisticians talk about "error" in accounting populations they often give the impression that they assume accounting errors to be a small set of homogeneous black balls mixed in with a large set of white balls.

When we note from Exhibit 3.11 the wide variety of error which exists in accounting populations it may be that this degree of variety affects the accuracy of the extrapolation of error value from the sample to the population.

Rates of compliance to monetary error

There has been some discussion in the auditing literature as to whether the proportion of compliance errors could be used to infer the proportion of monetary error in an accounting population. The so-called "smoke and fire" hypothesis:

The present study provides no support for this hypothesis. The compliance errors with no monetary affect were few and we could find no statistically significant relationship between the number of errors and the total value of error.

Rates of errors of overstatement to errors of understatement

An error of overstatement of book value is limited to the value of the item audited. An error of undersatement of book value can be of infinite value. This fact has been widely recognised as a problem when upper limits on error value are estimated from audit samples. However it has usually been argued that most errors are errors of overstatement and that most errors of understatement are small in value.

The findings of this study do not support this conclusion. Errors were only slightly weighted in favour of overstatements and the total value of each type of error was almost equal.

This finding conflicts with the findings of Ramage Krieger Spero (1979) study on the same point.

Certain statistical evaluation methods used to estimate upper limits on monetary error are only applicable to errors of over-statement of book value.

Frequency distribution of error values

The frequency distribution of the population of error values are set out in Exhibit 3.8.

We note the extreme skewness of this distribution, much affected by the two outliers. Therefore statistical estimation techniques based on the Normal Distribution would not be appropriate when inferring population measures from such a sample.

Tainting

The frequency distribution of tainting percentages is illustrated in Exhibit 3.10.

We note that the distribution is U shaped with relatively few observations between 10% and 99%.

Large taintings tend to occur in audit units of relatively small value Exhibit 3.10.

The difference between the pattern of tainting in debtors and inventory is statistically significant at the 95% level.

Those auditors using statistical estimation methods which are affected by the pattern of taintings may find these results of some interest.

Conclusions on the U.K. Study

A population of 323 errors discovered by external auditors in accounting populations of inventory and debtors (bills receivable) were classified and examined.

It was concluded that:

1. The value of total error corrected was not likely to be material and so the purpose of audit sampling would appear to be extrapolation rather than correction.

2. A wide variety of error type was noted.

3. A surprisingly low proportion of errors noted were compliance errors with no monetary effect.

4. Both in number and in total value overstatement of book value did not exceed understatement of book value by a large margin.

5. The frequency distribution of error values was highly skewed to the right.

6. The frequency distribution of error tainting percentages was U shaped. Few taintings occurred between 10% and 99%. Most large taintings occurred in audit units of low relative value. The difference between the distribution of taintings for inventory and debtors (bills receivable) was statistically significant.

3.7 Answers to questions set earlier on the nature of error in accounting populations.

We are now in a position to provide some tentative answers to the questions set in the earlier part of this chapter.

1. The mean rate of error.

This is usually low, below 1%, but large error rates are not rare. Rates exceeding 20% may be expected to occur in around 5% of audited populations.

Little evidence is available on the question of the constancy of the error rate. Gregory (1951) found that the error rate was not constant. Leslie and Anderson (1975), both experienced practitioners, appear to doubt the wisdom of depending on the constancy of previous years' error rates when designing an audit sample.

Audit	Mean	Standard deviation	Skewness	Kurtosis
		(a) Debtors		
6	.19	.11	2.16	16.28
11	.06	.33	1.30	4.16
15	.47	.33	.21	- 1.25
16	.18	.35	1.86	1.58
59	.59	.42	- .04	- 1.61
69	.22	.47	.17	.17
77	.49	.33	.30	- 1.37
80	.77	.35	-1.15	- .49
81	.93	.19	-3.61	11.93
91	.95	.15	-3.21	9.59
Median	.48	.33	.19	.88
		(b) Inventory		
13	-.70	3.64	-5.32	26.71
14	-.02	.34	-1.74	4.96
22	-.39	2.97	-6.42	41.40
23	-.13	1.98	-9.73	98.88
24	.08	.22	- .68	5.04
27	-.29	2.81	-8.44	73.62
32	-.02	.50	-4.18	26.09
33	-.29	2.17	-3.56	11.42
56	.03	.15	-2.29	9.33
60	.14	.50	.41	- .19
Median	-.08	1.24	-3.87	18.76

Exhibit 3.12 Some characteristics of error tainting distributions in debtors and inventory audits.

Source: JLN (1979)

We have no evidence as to whether the rate of error value varies from year to year. However, the PMM data strongly suggests that the mean rate varies between different types of accounting populations (see exhibit 3.12). The mean difference was significant at the 99% level.

The variance on the distribution of value errors in individual populations was extremely high. For example we calculate the median coefficient of variation to be:

Debtors	3.6
Inventory	29.7 (!)

4. <u>Are error values usually biased, usually skewed and do large understatements of error occur often or rarely?</u>

The PMM data suggests that since overstatements appear to occur much more often than understatements, error value is usually biased towards overstatement of recorded amount. The distribution of error values appear to be invariably highly skewed. In the PMM data the skewness of taintings varied from + 2.16 to - 9.73. The median tainting was + .19.

Understatements appear to occur quite frequently in inventory audits, fortunately large understatements are not common but they occur too frequently to be described as rare. Note the negative skewness attached to inventory taintings in Exhibit 3.12.

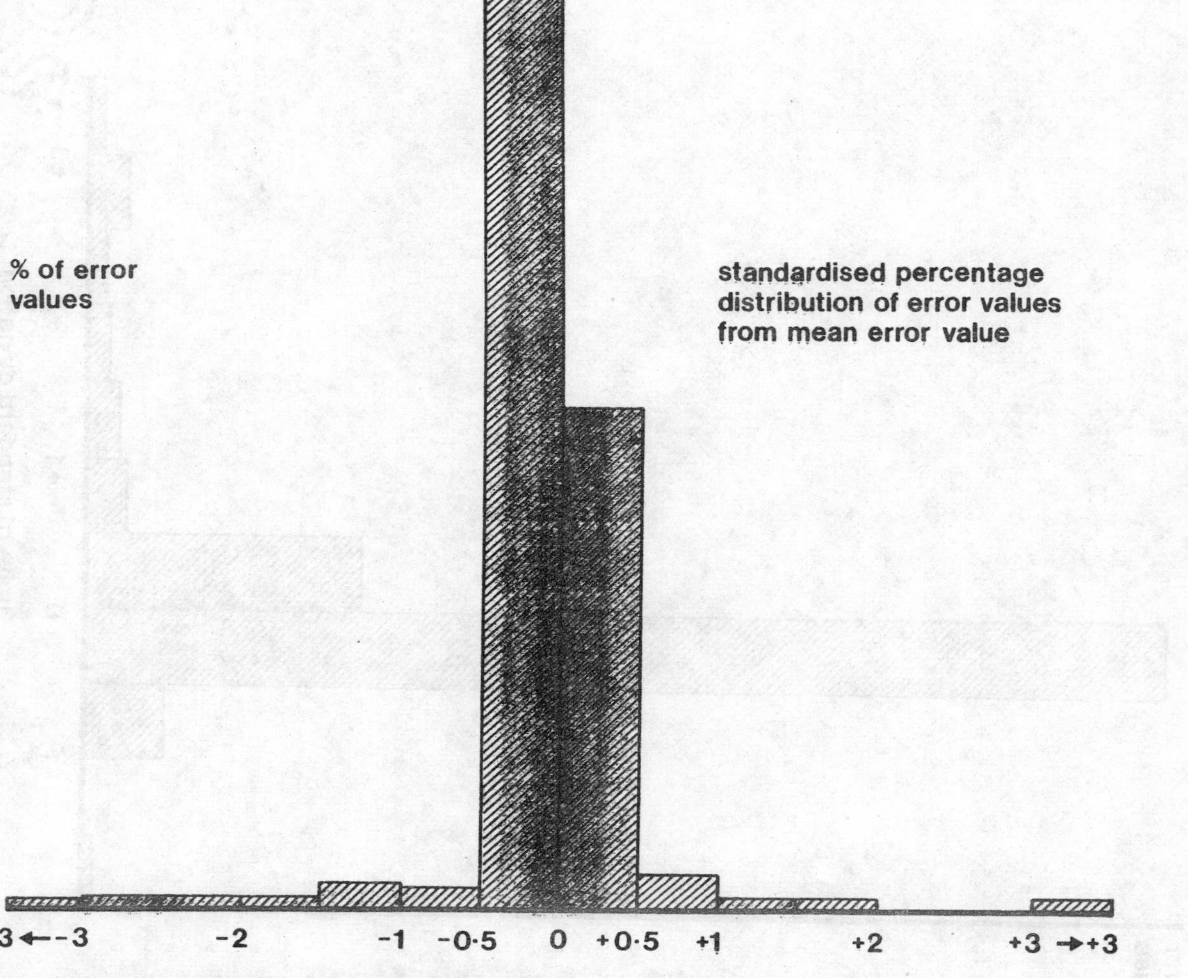

Exhibit 3.13a ERROR VALUES: INVENTORY
Adapted from JLN 1979

Note: 92 % of readings are within 0·5 standard deviations of mean error value

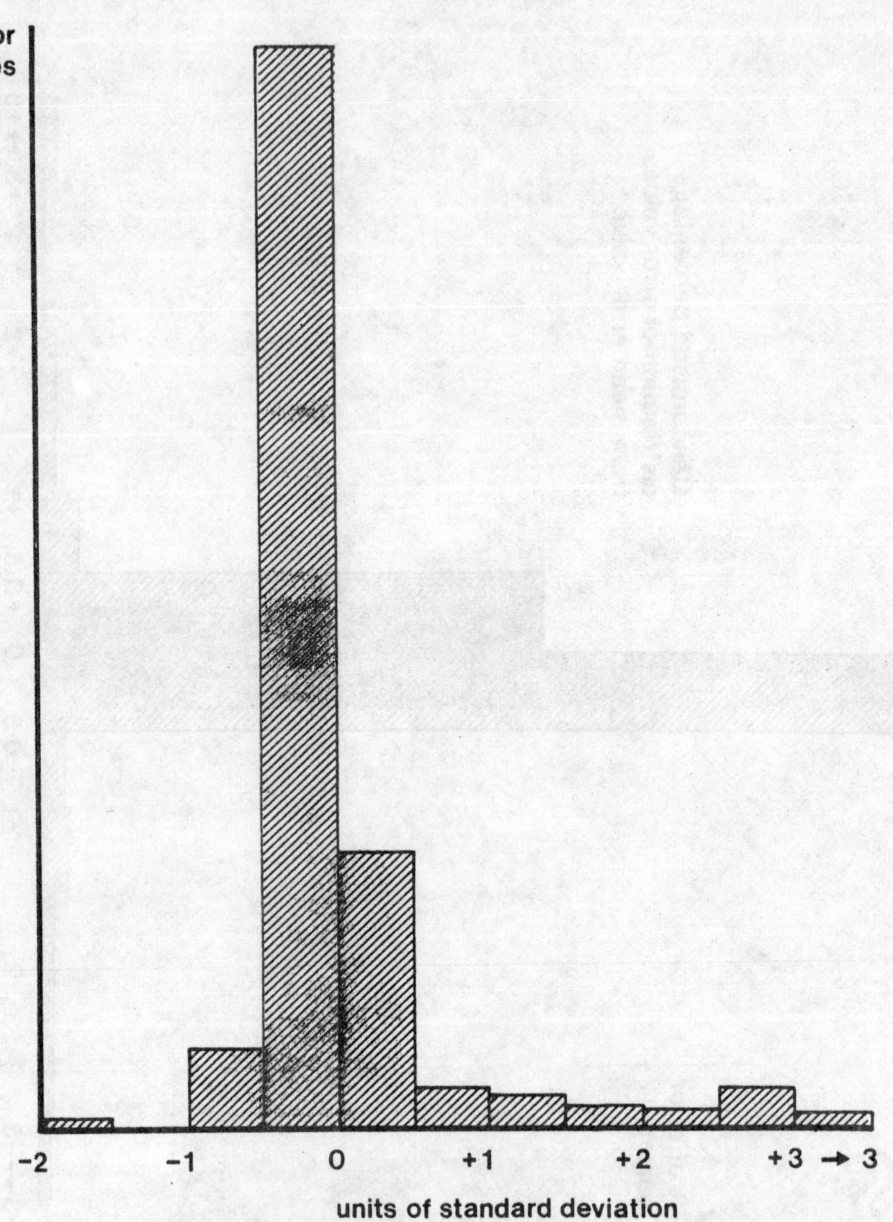

Exhibit 3.13b ERROR VALUES – DEBTORS
Standardised % distribution of error values from mean error value
Note: 86 % of readings are within 0·5 standard deviations of mean error value
Adapted from JLN (1979)

u

The standardised percentage (SP) distribution of error values calculated by JNL found a high proportion of values close to the mean. See exhibit 3.13 (a) and (b).

The SP distributions of inventory and debtors were not too dissimilar in shape, suggesting that all such distributions may approximate to a modal distribution.

Further research is needed to test this hypothesis.

5. <u>How frequently are zero values and negative values found in accounting populations?</u>

Little evidence is available to assist in answering this question. Only 1% of the 4038 errors in the PMM populations were excluded for being zero or negative. But this cannot be taken as a general guide since certain types of clients tend to have a high proportion of zero balances, i.e. hire purchase accounts.

6. <u>Are error values proportionate in size to the value of the item in which they occur?</u>

RKS, in their analysis of the PMM data, concluded that the answer to this question was 'yes'. JLN, on the other hand, found the variability of error value greater for large accounts but <u>not the magnitude</u> of the error. See Exhibit 3.14 which shows a mean R^2 of only 0.21 and 0.32.

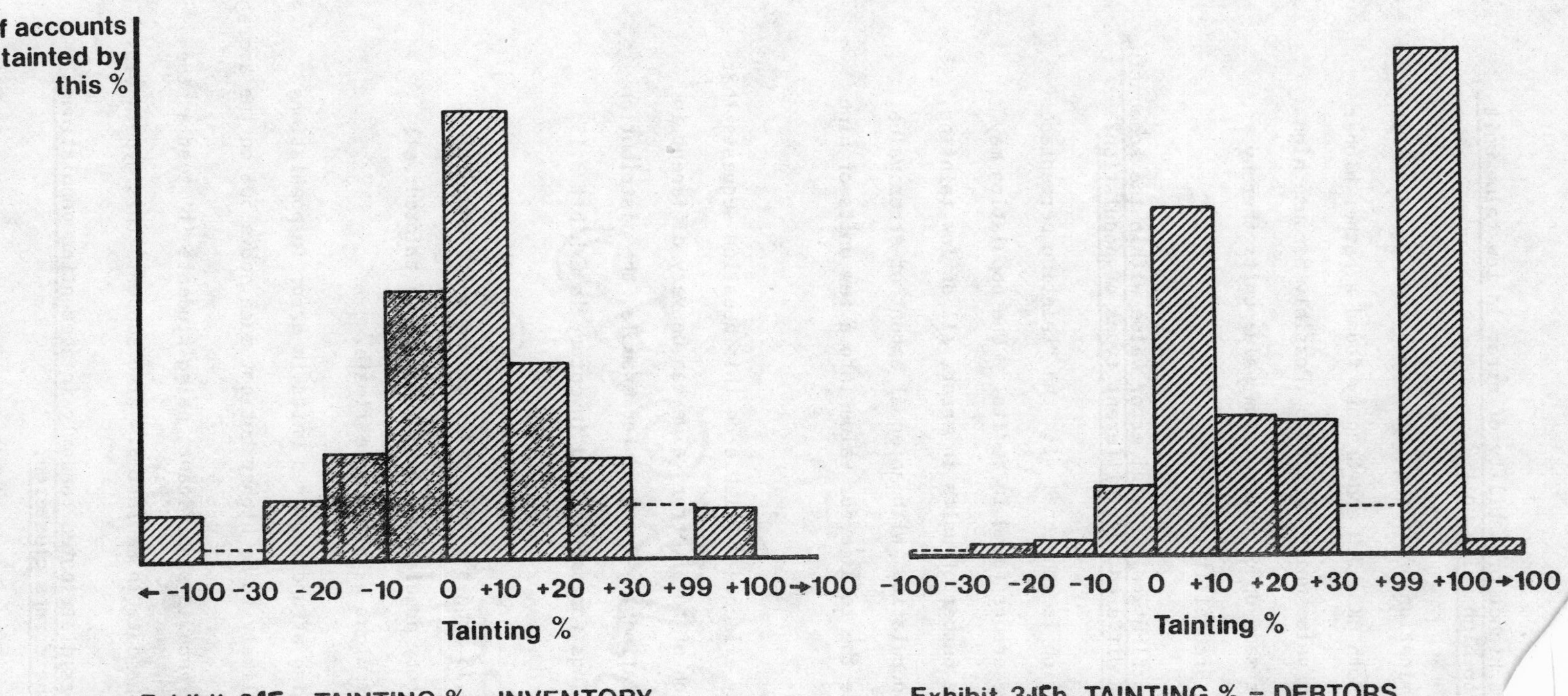

Exhibit 3·15a, TAINTING % – INVENTORY
Source: PMM study

Exhibit 3·15b, TAINTING % – DEBTORS
Source: PMM study

7. <u>Is there a higher probability of error in low value units compared to high value units?</u>

Every examination of this question has come to a negative conclusion. Both RKS, JLN and Gregory found a rather higher rate among high value units sampled. Possibly because high value units are made up of several low value units thereby increasing the probability of error.

8. <u>Is the distribution of total error value within the accounting population different for different types of population?</u>

This question is concerned with the "tainting percentages" attached to the errors found in audits. One population may contain a large number of units in error, all of low tainting, while another population, with an equal amount of error value, may concentrate the total error value into a few units of high error tainting.

The scanty evidence available on this question suggests that the distribution of total error value can be very different in different populations. Compare, for example, the distribution of error taintings for debtors and inventory in Exhibit 3.15 (a) and (b).

Exhibit 3.15 suggests that undiscovered units containing error in inventory populations are likely to be tainted well below 100%, 20% might be a better estimate.

The tainting of undiscovered units in error in populations of debtors is likely to be higher but not much above 50% on the average.

The difference is significant, using students 't' test at the 99% level. The data in Exhibit 3.4 also supports this hypothesis.

9. <u>Are errors distributed randomly in accounting populations or are the errors clustered?</u>

Of all the questions postulated this is the question least

likely to find an answer. The cost of auditing an entire accounting population would be prohibitive. A study of error frequency in cluster samples might provide a tentative answer.

Note that it is the distribution of error in the population under audit that is important, not the population of transactions generating the audit population. The sequence of the population under audit may be determined by alphabetical order while the sequence of the transaction population may be determined by time of arrival of input document.

If the audit sample is an unrestricted random sample the pattern of error is not important since the random selection, in effect, reshuffles the pack.

3.8. Conclusions

Audit sampling is mainly concerned with ensuring that the value of accidental operational error in accounting populations under audit is kept within acceptable bounds.

Audit sampling is also used to assess the degree of compliance errors occurring in the population under audit.

Surprisingly little is known about the pattern of error in accounting populations. Those few studies which have been published suggest that both error rates and error values are highly skew, that different types of accounting populations and different types of client have different error patterns, and that understatement errors and 100% errors are not rare events.

Two curious findings which conflict with conventional assumptions are that error rates tend to be higher in high value units compared to low value units and that the value of error is not strongly correlated to the value of the item in error.

Several of these findings could have a significant influence on the design and selection of sampling plans suited to audit work.

Notes

1. Assuming that both number and value are unbiased.

2. By "worst possible" is meant that distribution which would maximise beta risk given the sampling plan adopted.

3. See for example Gregory (1951) Neter and Loebbecke (1978) and a forthcoming paper by Johnson, Leitch and Neter based on the PMM error sample.

4. By understatement is meant that audited value (X_i) is less than the recorded value (Y_i).

5. In Cyert and Davidson (1957).

6. See, for example, Roberts (1978) p.122.
   Kaplan's adjustment to the DUS system to control alpha risk may cause a huge increase in sample size (2 to 3 times) because "a lower maximum relative error is seldom known".

7. Only 258 of the 298 errors had a value attached to them.

8. See for example CICA (1980), Akresh (1979) and McRae (1981).

Chapter 4.

A description of some scientific sampling methods currently used for external audit sampling.

4.1. Introduction

A wide range of scientific sampling methods have been used to select and evaluate external audit samples.

This chapter will describe those methods which have been used in practice. Audit sampling methods can be divided into two main categories. The first category is comprised of those sampling methods concerned with testing compliance with given accounting procedures. The second category is comprised of those sampling methods concerned with testing the accuracy of the value assigned to given accounting numbers.

The sampling methods used for compliance testing of procedures are mostly derived from the statistical theory based on attribute sampling. Examples of such methods are:

Estimation sampling of proportions.

Acceptance sampling.

Discovery sampling.

The acceptance sampling method may be single stage, multi stage or, at the limit, sequential.

The sampling methods used for substantive testing of value can be classified into two basic types. The first type is based on the theory of estimation sampling of variables, the second on a method of sampling called "probability proportionate to size" sampling (PPS).

The variables approach employ either mean per unit estimators or auxiliary estimators such as ratio estimation or difference estimation.

Stratification may be applied to improve the

sampling efficiency of each of these methods.

Probability proportionate to size (PPS) sampling is, technically, a form of attribute sampling. Several varieties of PPS have been developed and used in auditing. For example:

Monetary unit sampling (MUS)
Dollar unit sampling (DUS)
Comulative monetary amount sampling (CMA)
Combined attributes and variables sampling (CAV)

We shall call this sampling method monetary unit sampling (MUS) unless we are referring specifically to one of the variants listed above.

The remainder of this chapter is devoted to a description of each of these methods as they are applied to external auditing. The statistical formula underlying the method will be set down but not proved.

The proof, in most cases, is available in standard statistical textbooks. When this is not the case, as with MUS, we will provide an extensive set of references.

4.2. Sampling methods for estimating error rates in accounting populations under audit.

4.2.1. Estimation sampling of proportions.

If an external auditor wishes to use SS methods to assist in compliance testing of procedures, he can either estimate an upper bound on the error rate in the accounting population under audit or he can test the hypothesis that the error rate is more than some given %.

At this stage in the audit process the value of error need not be considered, only the rate of error. The rate of error represents that proportion of the population under audit which fail to comply with the accounting procedures laid down in the procedures manual or fail to conform to accepted accounting practice.

The sampling process is as follows. First the auditor lists the various sets of procedures under audit, second he draws a random sample from some or all of these populations of procedures, finally he audits this sample.

The auditor must have a prior notion as to what he considers to be an acceptable degree of error. In effect he samples the population of accounting procedures to test the hypothesis that the actual error rate in the population under audit exceeds this acceptable degree of error.

If he is using estimation sampling of proportions he must decide on the following factors prior to sampling.

(a) The confidence level he requires in his inference from the sample.

(b) The expected error rate.

(c) The maximum acceptable upper limit on the estimate of the error rate (the upper bound on the confidence interval).

The choice of confidence level is almost invariably a subjective estimate based on the auditors prior knowledge about the quality of the accounting procedures in this particular audit. Kinney (1977) has suggested a rather more scientific approach by using the results of analytical review, measured by regression analysis, to determine the confidence level required for sampling.

The estimate of the expected error rate will be based on either the previous years audit papers or on error rates encountered in similar clients audited by the auditor.

The upper error limit (UEL) is another subjective estimate. The audit profession could indicate standards for UEL's under specific situations but they have not done so. Each auditor must make up his own mind as to what is a suitable UEL for each particular audit.

As we noted above the auditor is, in effect, testing the hypothesis that the actual error rate in the population exceeds the acceptable rate. If the sample estimate of the error rate is less than the UEL he rejects the hypothesis that the population error rate exceeds the acceptable rate. He therefore accepts the population as being of good quality.

Note that the confidence level and UEL are decided <u>subjectively</u> by the auditor. These two decisions are based on training and experience <u>not</u> on a universally accepted audit standard. Thus even when SS is used in auditing, inconsistent standards of audit can still apply <u>between</u> auditors, even if a more consistent standard is applied by any given auditor, or within any given audit firm.

<u>The formulae for calculating sample size in estimation sampling of proportions</u>.

The formula used to calculate the sample size required is estimation sampling of proportions is as follows (1).

$$n = \frac{p(1-p)}{(e/f)^2 + p(1-p)/N}$$

when n = required sample size, p = expected proportion of sample with condition, e = required precision limit, ± e%, f = factor decided by confidence level (90% = 1.64, 95% = 1.96, 99% = 2.58, 99.7% = 2.97), N = population size.

When in doubt it is advisable to make a conservative estimate of p, the estimated error proportion. In this case a conservative estimate means a low, rather than a high error estimate.

The standard error of the sample estimate is given by the formula:

$$s = \left(\frac{p(1-p)}{n}\right)^{1/2}\left(1 - \frac{n}{N}\right)^{1/2}$$

where s = standard error of a proportion, p = percentage of sample with condition, n = sample size, N = population size.
This formula becomes increasingly inaccurate as p approaches 0%. Below about 3% it is advisable to use corrected tables such as Arkin (1963), Table F.

In practice, as we stated above, e and f are determined subjectively by the auditor, p is estimated, and N is often assumed to be so large as for all practical purposes to be infinite. In consequence of this last assumption $p(1-p)/N$ and $\left(1-\frac{n}{N}\right)^{\frac{1}{2}}$ ~ the population correction factors ~ are eliminated from the equations.

The ratio p, that is the estimated error rate in the population, has a large affect on the standard error as p becomes a very small fraction, say below 3%. See Arkin (1974) Table F for a method of handling this problem.

For a fixed sample size the coefficient of variation of the estimate increases rapidly as p falls below 3%. Thus very large samples are needed for <u>precise</u> estimates of error rate below 3%. As Cochrane has written,

> "simple random sampling.....is an expensive method of estimating the total number of units of a scarce type." (2)

Since accounting errors are usually scarce, it is much more economical for an auditor to test that the proportion of error does not <u>exceed</u> some given upper error limit rather than to try and measure the precise number of errors in a given accounting population, under audit.

The auditor should always aim to collect only the <u>minimum</u> information needed to satisfy his audit objectives.

For this reason acceptance sampling would seem to be better suited to compliance testing rather than estimation sampling of proportions.

4.2.2. <u>Acceptance sampling</u>.

Once an auditor has drawn his sample, using estimation sampling of proportions, he calculates a ratio. This ratio is $e \div n$, where e represents the number of errors in the sample and n the sample size. He then extrapolates this error ratio to the population under audit.

The first auditor to make a serious study of applying SS to external auditing (Vance 1950) noted that such an approach differed from conventional audit procedure. Under conventional audit procedure the auditor draws his sample and counts the number of errors. The absolute number of errors is noted. This figure will, of course, tend to vary with sample size, but under traditional audit practice the <u>ratio</u> of error is not calculated. The auditors judgement as to the quality of accounting procedures under audit is influenced by the number and <u>quality</u> of the individual errors discovered.

This fact suggested to Vance that some form of acceptance sampling was better suited to compliance testing rather than estimation sampling of proportions.

The celebrated book by Vance and Neter (1956), which launched SS as an audit tool, deals extensively with various forms of acceptance sampling.

Acceptance sampling was designed by industrial quality control experts to control the quality of inventory received in batches. A batch of goods is only accepted if it contains less than a given number of rejects. Since it is not economical to test every item in the batch the accept/reject decision is based on a statistical evaluation technique.

The tester draws a sample of a given size from the batch, he identifies any rejects, he counts the number of rejects and if these are equal to a less than some given number, c, he accepts the batch. If the number of rejects exceed the given number, c, he rejects the batch.

The sample size in acceptance sampling is calculated from the four parameters.

AER = acceptable error rate.

UER = unacceptable error rate

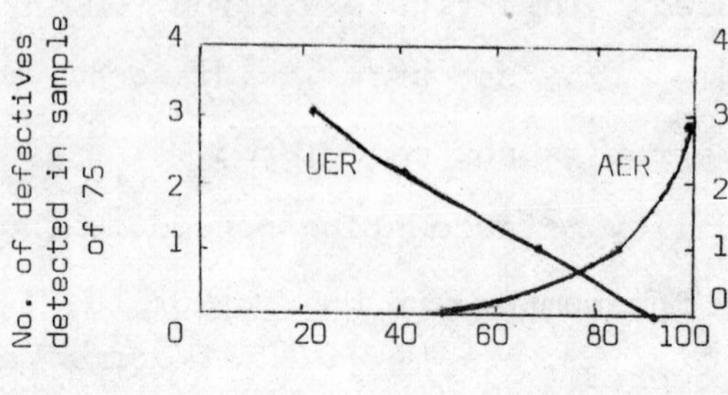

The graph shows the level of confidence of rejecting batches with 3% error or above (UER curve) along with the level of confidence of not rejecting batches with error rates as low as 1% (AER curve).

Population	1000
UER	3% (Beta risk)
AER	1% (Alpha risk)
Sample size	75

Notice that the levels of confidence of the two curves run counter to one another.

Exhibit 4.1. The chart illustrates the problem of balancing alpha against beta risk in acceptance sampling.

a = alpha risk i.e. risk of rejecting acceptable batch.

b = beta risk i.e. risk of accepting unacceptable batch.

Extensive acceptance sampling tables have been calculated and published to assist the tester in industrial quality control applications. See for example Dodge and Romig (1944). These tables are not suited to external auditing. However, fortunately, a set of acceptance sampling tables suited to external auditing has been produced by Arkin (1974) - appendix K p 473. Finley (1978) suggests that an auditor should calculate his own tables (P.34).

The major advantage acceptance sampling enjoys over estimation sampling of proportions, is that it allows the auditor to balance out alpha risk against beta risk. As was explained in Chapter two alpha risk is the risk that an auditor might reject an account he should have accepted, while beta risk is the risk that an auditor might accept an account he should have rejected.

Traditionally an external auditor lays more emphasis on minimising beta risk rather than alpha risk. Acceptance sampling allows the auditor to <u>choose</u> a sampling plan which provides adequate cover against both alpha and beta risk. The technique is illustrated in Exhibit (4.1).

An even tighter control system can be introduced if the auditor checks out every unit in the rejected batches. Under this approach whatever the error rate in the batch may be, the maximum error rate which can slip through the test, if all rejected batches are inspected 100%, is limited to x% for a batch error rate of Y%. Thus the auditor can limit the maximum average error rate remaining in the population after audit to a given %.

Kriens and Dekkers (1980) of the Netherlands seem to have made this approach the basis of their SS audit system (3).

We believe this system to be too expensive for normal audit work.

<u>The formula for calculating acceptance samples.</u>

To calculate acceptance sampling tables one should, ideally, use the hypergeometric distribution formula.

This is,

$$r = \left(\frac{a!}{b!(a-b)!}\right)\left(\frac{(N-a)!}{(n-b)!\{(N-a)-(n-b)\}!}\right)\Big/\left(\frac{N!}{n!(N-n)!}\right)$$

r = probability of this situation occurring, a = number of units in population with given condition, b = number of units in sample with given condition, n = number of units in the sample, N = number of units in the population.

When the population is infinite, i.e. all samples drawn are replaced (4) one can use the simpler binomial formula.

$$r = \frac{n!}{b!(n-b)!} \; p^{b}(1-p)^{n-b}$$

were r = probability of this situation occurring, n = number of units in the sample, b = number of units in sample with specified condition, p = proportion of population with specified condition.

Most auditing applications use such large sample sizes that the above computations would be tedious unless sampling tables were made available to the auditor. Arkin (1974) provides an extensive set of tables to assist auditors in calculating sample plans for acceptance and discovery sampling.

4.2.3. Multiple sampling.

When using the methods discussed above, if the number of errors discovered in the sample exceeds the acceptable number the population is rejected, if the number of errors discovered equals or is less than the acceptable number of errors the population is accepted.

It is sometimes more economical to allow three decisions rather than two, namely accept, reject or take a bigger sample. Populations of very high or low quality are quickly accepted or rejected, populations in between are subjected to more sampling. This method is called multiple sampling.

This method is intuitively appealing to the auditor and is economical on sample size, but it can be rather complicated in practice and for this reason seems to have received limited application in external auditing. The economy in average sample size arising from the rapid identification of very good or very bad populations may be cancelled out by the cost of organising the second sample.

Several multiple acceptance sampling plans have been designed and applied (See Roberts 1978 pages 57-61). These plans cannot be constructed directly from attribute tables but must be designed specifically for the required plan. The reason for this is that the intermediate result at each stage of the sampling has an effect on determining the total sample size. (See Wilburn (1968))

A table for constructing double sampling plans, devised by Paul Peach, is attached as an appendix to this chapter.

4.2.4. Sequential sampling.

Sequential sampling carries multiple sampling to the limit since the auditor tests on each sequential unit audited to see whether the population should be accepted or rejected or whether the sampling should be continued.

If the actual population error rate is very high then this will show up within a few observations. Suppose, for instance that out of the first ten items sampled at random, four are found to be in error. This provides sufficient evidence to suggest that the population error rate exceeds, say, 5% and there is clearly no point in continuing with further testing. This process is called sequential acceptance sampling. Random units are selected and tested one at a time. On each test one of the following decisions is reached:

(a) There is sufficient evidence available to "accept".

(b) There is sufficient evidence available to "reject".

(c) Insufficient evidence exists to reach either an "accept" or a "reject" decision, so a further unit is selected at random and tested.

Diagrammatically this may be expressed as

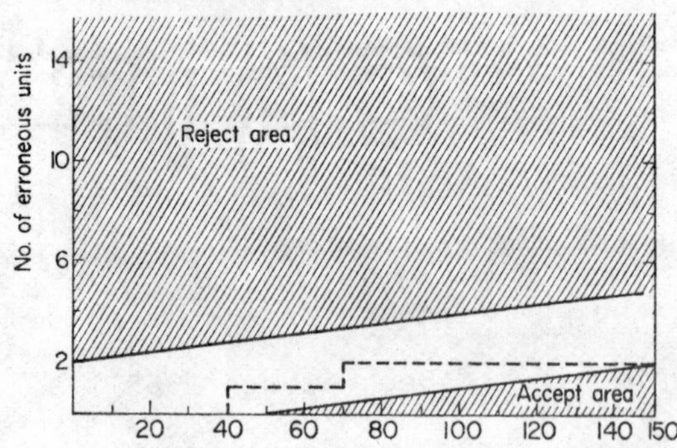

No. of units audited

Sequential sampling acceptance table. The 40th and 70th units contain an error but the batch is eventually accepted after 150 units have been audited.

It can be shown mathematically that the average sample size is smaller for sequential than for normal acceptance sampling. However the testing of the population after every unit sampled is costly in audit time and this may exceed the benefit of the smaller average sample size.

The method would seem to be rarely used in practice in external auditing. (5). Formulae for calculating sequential sampling plans are outlined in Vance and Neter (1956).

4.2.5. Discovery sampling.

Acceptance sampling allows the auditor to balance out alpha and beta risk. If, however, the auditor is prepared to ignore alpha risk, that is if he is prepared to ignore the risk of rejecting acceptable populations (or overauditing) he can use a form of acceptance sampling called discovery sampling.

This method of audit sampling was first suggested by H. Arkin (1961). It is that specific form of acceptance sampling where, c, the maximum acceptable number of errors is always equal to zero. If any procedural errors are discovered in the audit sample the population is immediately rejected. The method may be economical when auditing groups of accounting populations which are subject to a wide variation in error rate. If only a few populations are expected to be of poor quality the auditor can identify these at low sample cost.

The trouble with discovery sampling is that alpha risk is ignored. Under normal audit conditions it is likely that many accounting populations which should be accepted are rejected as being of poor quality.

Discovery sampling tables are designed to calculate that sample size which will discover at least one example of an error if the population error rate exceeds some specified minimum % (at a given confidence level). It is designed to detect at least one example rather than to estimate a rate of occurrence. It may be useful where almost no degree of error is tolerable in an important audit area.

The formula for calculating discovery sampling tables is provided in Arkin (1974).

4.3. Sampling methods for estimating the total value of error in accounting populations under audit.

Estimation sampling of variables.

All of the sampling methods discussed above are designed to estimate either an error rate or an upper limit on an error rateor whether an error rate exceeds some upper limit. They are designed to test for the degree of compliance with stated accounting procedures. These sampling methods test for error rates not error values, they are suited to testing stage two of the audit plan suggested in Chapter two above.

If an auditor wishes to estimate error value, or a limit on error value, he must employ some other form of sampling design.

Three methods of sampling have been employed to estimate error value. These methods are:-

1. Mean per unit sampling - with or without stratification.
2. Auxiliary estimator sampling - with or without stratification.

3. Probability proportionate to size sampling (6).

4.3.1. Mean per unit sampling

If an accounting population has a low value variance, which is unusual, or if the individual sampling units are not recorded, simple mean per unit estimation sampling of variables can be employed to estimate the audited value of the population.

The statistics needed to calculate audit sample size in mean per unit sampling are as follows:

(a) sample mean of audited amounts.

(b) estimated standard deviation of audited amounts.

(c) estimated standard error of mean estimator.

The various formulae are set out in Exhibit (4.2).

The problem with calculating an unstratified mean per unit estimate is that most accounting populations are highly skewed as to value (7). The population of error values is also highly skewed, as demonstrated in Chapter three. As we will show in Chapter eight this degree of skewness generates audit samples of a size much larger than those traditionally drawn in audits in the U.K.

The solution to this problem is to apply some form of stratification to the accounting population under audit.

Unstratified Mean Estimation

Formula	*Description*
$\bar{x} = \frac{\sum x_i}{n}$	Sample mean of audited amounts
$\hat{X}_M = N\bar{x}$	Unstratified mean estimator of total audited amount
$S_x = \sqrt{\frac{\sum x_j^2 - n\bar{x}^2}{n-1}}$	Estimated standard deviation of audited amounts
$\hat{\sigma}(\hat{X}_M) = \frac{NS_x\sqrt{1-n/N}}{\sqrt{n}}$	Estimated standard error of unstratified mean estimator
$n = \frac{N^2U_R{}^2S_x{}^2}{A^2 + NU_R{}^2S_x{}^2}$	Sample size formula for unstratified mean estimator
$A'_M = \frac{NU_RS_x\sqrt{1-n/N}}{\sqrt{n}}$	Achieved precision of unstratified mean estimator

Where

$\bar{x}$ = sample mean
x = value of unit sampled
n = number of units in sample
$\hat{X}$ = Unstratified mean estimator
N = Number of units in the population
S = Estimated standard error of mean estimator
U = Reliability factor (confidence) required.
A = Absolute precision required by auditor.

Exhibit (4.2) Formulae for calculating mean per unit estimates (unstratified). Source: Roberts (1978) p.237.

4.3.2. Mean per unit with stratification

Mean per unit can be used more effectively if some form of stratification is applied. The variability of value within each stratum is reduced and so the required sample size is reduced for any given level of precision and confidence.

Normally the top value stratum is sampled 100% and so the problem is limited to deciding on stratum boundaries for lower value units and allocating the total audit sample between the various strata.

Several methods have been suggested for locating stratum boundaries. Two popular methods used are as follows:

1. Allocate boundaries in a way that each stratum contains approximately the same total value.

2. Decide on the required number of strata and divide the <u>cumulative</u> square root of the number of units in

each stratum by this number. Allocate stratum boundaries so that each stratum has approximately this cumulative square root frequency. The Price Waterhouse audit computer package uses this particular system.

Once the stratum boundaries are assigned the next step is to allocate the total audit sample between the strata.

Three methods commonly used are as follows.

1. Allocate the total sample between each stratum in proportion to the total recorded amount of each stratum.

2. Allocate the total sample in proportion to n x s, where n represents the number of units in each stratum and s represents the standard deviation of the stratum. This method minimises the standard error of the standard deviation, it is sometimes called the Neyman method.

3. Allocate the total sample between each stratum in proportion to n x 1. Where n is as before and 1 represents the largest recorded amount in each stratum.

The formulae for calculating stratified mean per unit estimators are set out in Exhibit 4.3.

4.3.3. Auxiliary estimators - ratio and difference estimates.

It is sometimes argued that mean per unit sampling is aimed at the wrong target. Auditors are specifically concerned with differences between audited and recorded amounts. Mean per unit estimation concentrates on totals rather than differences.

There are however statistical sampling techniques available which are designed to estimate differences and ratios.

Stratified Mean Estimation

Formula	Description
$\hat{X}_{MS} = \sum N_i \hat{x}_i$	Stratified mean estimator of total audited amount.
$\hat{\sigma}(\hat{X}_{MS}) = \sqrt{\sum N_i(N_i - n_i) \frac{S_{xi}^2}{n_i}}$	Estimated standard error of the stratified mean estimator
$n = \frac{U_R^2(\sum N_i \sigma_{Y_i})^2}{A^2 + U_R^2 \sum N_i \sigma_{Y_i}^2}$	Sample size formula for the stratified mean estimator using optimal allocation.
$n = \frac{U_R^2 Y \sum N_i^2 \frac{\sigma_{Yi}^2}{Y_i}}{A^2 + U_R^2 \sum N_i \sigma_{Yi}^2}$	Sample size formula for the stratified mean estimator using PRA allocation (proportional to the recorded amount)
$A'_{MS} = U_R \sqrt{\sum N_i(N_i - n_i) \frac{S_{xi}^2}{n_i}}$	Achieved precision of the stratified mean estimator.

Where definition of symbols are as given in exhibit (4.2) on page (112) except that:

Y = total recorded amount in the population.

σ_{Y_i} = standard deviation of recorded amounts in the i stratum.

i = stratum

Exhibit (4.3) Formulae for calculating stratified mean per unit estimates (From Roberts (1978) p. 239)

4.3.4. Difference estimation

This method lists the differences, plus or minus, between individual recorded values and audited values.

The average difference in the sample is then multiplied by the number of units in the population to arrive at an estimate of the total difference. An estimate of the total audited amount is arrived at by adjusting the total recorded amount for the total estimated difference.

The standard error of the estimate of difference equals the standard deviation of the sampling distribution. Unfortunately the rarity of value error in most accounting populations plus the high skewness of value differences limits the use of this sampling method in practice. The size of audit sample required to reduce the standard error of the estimate to reasonable proportions is usually much too high.

The formulae for calculating unstratified difference estimates are set out in Exhibit (4.4).

This sampling method is at its most efficient when the differences are approximately equal in value no matter what the value of the recorded amount in error.

4.3.5. Ratio estimation

This method of sampling is very similar in approach to difference estimation. The method estimates the ratio of the recorded amount to the audited amount.

If the proportion of error is low, one sided or highly skewed as to value, we find similar problems to those listed under difference estimation. Under these conditions the sample distribution is unlikely to

Unstratified Difference Estimation

Formula	Description
$\hat{D} = N\bar{d}$	Unstratified difference estimator of total difference
$\hat{X}_D = Y + \hat{D}$	Unstratified difference estimator of total audited amount.
$S_D = \sqrt{\frac{\sum d_i^2 - n\bar{d}^2}{n-1}}$	Estimated standard deviation of difference amounts.
$\hat{\sigma}(\hat{D}) = \frac{NS_D\sqrt{1-n/N}}{\sqrt{n}}$	Estimated standard error of unstratified difference estimator
$n = \frac{N^2U_R{}^2S_D{}^2}{A^2 + NU_R{}^2S_D{}^2}$	Sample size formula for unstratified difference estimator.
$\sqrt{\hat{p}_D\sigma_Y{}^2 + \hat{p}_D(1-\hat{p}_D)\bar{Y}^2}$	Approximate expected value of standard deviation of differences when all differences are 100 per cent overstatements
$A'_D = \frac{NU_RS_D\sqrt{1-n/N}}{\sqrt{n}}$	Achieved precision of unstratified difference estimator
$\hat{\sigma}_D = \sqrt{P_U(m)S_D{}^2(m) + P_U(m)(1-P_U(m))\bar{d}_m{}^2}$	Estimated standard deviation of difference amounts (approximation)

Where definition of symbols are as given in exhibits (4.2) and (4.3) except for:

d = difference
D = difference estimator
$\bar{d}$ = sample mean of the differences.
p = proportion of non-zero differences.
$\bar{Y}$ = mean recorded amount
m = number of differences observed

Exhibit (4.4) Formulae for calculating unstratified differences estimates. (From Roberts (1978) p. 237)

approximate to a normal distribution.

Cochrane(1963) states that the sample distribution will only approximate to normality if the sample is of sufficient size to make the coefficientof variation of recorded values divided by the square root of the number of units in the sample, less than 10%. Thus the sample size must be 100 times the square of the coefficient of variation of recorded amounts. The C.of V. in accounting population is usually between 2 and 5. Thus this suggests audit samples of 400 to 2500 units. This sample size is much too large for conventional auditing.

We will also find that many audit samples contain an insufficient number of differences - say 30 - or that the direction of difference is very uneven. The formula for calculating ratio estimates are presented in exhibit (4.5).

The ratio estimation sampling method is at its most efficient when the ratio of differences to recorded amounts are relatively constant.

4.3.6. Ratio and difference estimate with stratification

The sampling procedures using difference and ratio estimate tend to generate audit sample sizes which are too large for conventional auditing.

It is possible that the required sample size can be sufficiently reduced for the method to become economically viable if stratification is used.

Unstratified Ratio Estimation

Formula	*Description*
$R = \frac{\sum x_i}{\sum y_i} = 1 + \frac{\sum d_i}{\sum y_i}$	Estimated ratio of audited amount to recorded amount
$\hat{X}_R = \hat{R}Y$	Unstratified ratio estimator of total audited amount
$S_R = \sqrt{\frac{\sum x_i^2 + \hat{R}^2 \sum y_i^2 - 2\hat{R} \sum x_i y_i}{n-1}}$ $= \sqrt{\frac{\sum d_i^2 + (\hat{R}-1)^2 \sum y_i^2 - 2(\hat{R}-1) \sum d_i y_i}{n-1}}$	Estimated standard deviation of ratios
$\hat{\sigma}(\hat{X}_R) = \frac{NS_R \sqrt{1 - n/N}}{\sqrt{n}}$	Estimated standard error of unstratified ratio estimator
$n = \frac{N^2 U_R^2 S_R^2}{A^2 + NU_R^2 S_R^2}$	Sample size formula for unstratified ratio estimator
$\sqrt{\hat{p}_D(1-\hat{p}_D)(\sigma_Y^2 + \bar{Y}^2)}$	Approximate expected value of standard deviation of ratios when all differences are 100 percent overstatements
$A'_R = \frac{NU_R S_R \sqrt{1 - n/N}}{\sqrt{n}}$	Achieved precision of unstratified ratio estimation
$\hat{\sigma}_R = \sqrt{P_U(m) S_R^2(m)}$	Estimated standard deviation of ratios (approximation)

Where definition of symbols are as given in exhibits (4.2), (4.3) and (4.4) except for,

R = Estimated ratio of audited amount to recorded amount
x = audited value
y = recorded value

Exhibit (4 .5) Formulae for calculating unstratified ratio estimates (From Roberts (1978) p. 238)

Again the sampling distribution will only approximate to normality if the proportion of non-zero differences exceeds about 5% (small value differences) or about 30% (large value differences). See Neter and Loebbecke (1975).

The auditor will use the standard deviation of recorded amounts within each stratum to calculate the sample size as set out above, under mean per unit stratification.

So long as the correlation between audited and recorded amounts within each stratum exceeds one half, the sample size estimate is conservative.

Roberts (1978) writes that,

> "basing the total sample size on the standard deviation of recorded amounts and allocating the sample in proportion to the stratum total recorded amounts is a very good procedure". (Page 105)

When

1. Stratum boundaries are set to include equal recorded amounts.
2. Most differences are overstatement.

The sampling system works best for difference estimates when several differences occur in each stratum and larger differences occur in higher value strata.

Stratification using combined ratio estimates is similar in procedure to difference estimation except that the ratio estimate enjoys the advantage that the constraints imposed on number and distribution of differences are less stringent.

The formulae for calculating stratified difference

estimates and combined ratio estimates are set out in exhibits 4.6. and 4.7.

4.3.7 Regression estimates

It has been suggested that regression estimates could also be used for audit sampling. The method is described in Roberts (1978).

Our survey of audit practice discovered no accounting firm in the U.K. who used this method for audit sampling. For this reason we have not included it in our review.

Regression analysis is used by D.H.S. as part of their STAR system of analytical review.

4.3.8. Monetary unit sampling

The value sampling methods discussed to date suffer from at least one of two possible defects. First, if no stratification is applied the sample size is usually very large. Second if stratification is applied the method becomes complicated.

Monetary unit sampling (MUS) was developed to solve these two problems simultaneously. MUS, it is claimed, can satisfy normal requirements on accuracy at a reasonable sample size and yet remain relatively simple in application.

MUS sampling procedures are designed to follow traditional audit sampling procedures as closely as possible.

Monetary unit sampling is a form of attribute sampling which can be used to estimate a variable i.e. error value. (The difference between the recorded value and the audited value).

Stratified Difference Estimation.

Formula	Description
$\hat{D}_S = \sum N_i \bar{d}_i$	Stratified difference estimator of total difference.
$\hat{X}_{DS} = Y + \hat{D}_S$	Stratified difference estimator of total audited amount.
$\hat{\sigma}(\hat{D}_S) = \sqrt{\sum N_i(N_i - n_i)\,\frac{S_{Di}^2}{n}}$	Estimated standard error of stratified difference estimator.
$n = \frac{U_R^2(\sum N_i S_{Di})^2}{A^2 + U_R \sum N_i S_{Di}^2}$	Sample size formula for the stratified difference estimator using optimal allocation.
$n = \frac{U_R^2 Y \sum N_i^2 \frac{S_{Di}^2}{Y_i}}{A^2 + U_R^2 \sum N_i S_{Di}^2}$	Sample size formula for the stratified difference estimator using PRA allocation.
$A'_{DS} = U_R \sqrt{\sum N_i(N_i - n_i)\,\frac{S_{Di}^2}{n_i}}$	Achieved precision of the stratified difference estimator.
$\hat{\sigma}_D(i) = \sqrt{P_U(m)S_D^2(m_i) + P_U(m)(1 - P_U(m))\bar{d}_{m_i}^2}$	Estimated standard deviation of difference amounts in the ith stratum (approximation)

Where definition of symbols are as given in exhibits (4.2) and (4.4).

Exhibit 4.6. Formulae for calculating stratified difference estimates (from Roberts (1978) page 240).

Combined Ratio Estimation

Formula	Description
$\hat{R}_C = \frac{\sum N_i \bar{x}_i}{\sum N_i \bar{y}_i} \qquad = 1.0 + \frac{\sum N_i \bar{d}_i}{\sum N_i \bar{y}_i}$	Combined ratio estimator of population ratio of audited to recorded amount.
$\hat{X}_{RC} = \hat{R}_C Y$	Combined ratio estimator of total audited amount
$\hat{\sigma}(\hat{X}_{RC}) = \sqrt{\sum N_i (N_i - n_i) \frac{S^2_{RCi}}{n_i}}$	Estimated standard error of the combined ratio estimator
$S_{RCi} = \sqrt{\frac{\sum x_{ij}^2 + \hat{R}_C^2 \sum y_{ij}^2 - 2\hat{R}_C \sum x_{ij} y_{ij}}{n_i - 1}}$	Estimated standard deviation of ratio in ith stratum
$n = \frac{U_R^2 (\sum N_i S_{RCi})^2}{A^2 + U_R^2 \sum N_i S^2_{RCi}}$	Sample size formula for the combined ratio estimator using optimal allocation
$n = \frac{U_R^2 Y \sum \frac{N_i^2 S^2_{RCi}}{Y_i}}{A^2 + U_R^2 \sum N_i S^2_{RCi}}$	Sample size formula for the combined ratio estimator using PRA allocation
$A'_{RC} = U_R \sqrt{\sum N_i (N_i - n_i) \frac{S^2_{RCi}}{n_i}}$	Achieved precision of the combined ratio estimator

Where definition of symbols are given in exhibits (4.2) and (4.5)

Exhibit 4.7. Formulae for calculating combined ratio estimate. (From Roberts (1978) page 240)

The method has been identified by several names i.e.

Monetary unit sampling (MUS)
Dollar unit sampling (DUS)
Cumulative monetary amount sampling (CMA)
Combined attributes and variables sampling (CAV)
Probability proportionate to size sampling (PPS)

These methods are not identical in operation but they adopt a similar approach. Statistisians use the term PPS for this method. We shall use the term monetary unit sampling (MUS) to describe this particular approach to audit sampling.

Applying the MUS method

As we noted above several versions of the MUS method are in use. The version we are about to describe is a simplified form of the DUS method described is Leslie, Teitlebaum, Anderson (1980). We believe that this simplified version includes all the key elements of MUS but excludes the ideosyncracies of any specific version.

The steps required to implement MUS as an audit sampling method are as follows:

1. Decide on reliability$^{(R)}$ and materiality$^{(£M)}$ for this particular audit

Comment

Reliability is the level of confidence the auditor requires in his inference from the sample. This will normally vary between 80% and 95% depending upon his prior information about the quality of the particular accounting population under audit. Reliability as low as 39% (i.e. beta risk of 61%) is provided in some MUS sampling tables! (10)

The absolute value of materiality will normally be decided independently of the population under audit. Materiality is usually derived from the profit figure in the accounts under audit. Thus materiality could constitute any fraction of the value of the population under audit, it could even exceed the total value of this population!

2. Decide on an acceptable upper error limit (UEL) on the estimate of error value in the population under audit assuming that no monetary errors are discovered in the audit sample.

Comment

The UEL must be less than materiality (£M) otherwise the MUS sample becomes a discovery sample. If the sample were a discovery sample this would imply immediate rejection of the population if even one monetary error were found. This limitation can be put another way by saying that alpha risk is too high if £M = U. Kaplan (1975) provides a useful table which illustrates this point. Many writers on MUS have suggested that the value of the UEL should be fixed at 50% of the value of materiality under normal audit conditions. See Leslie etc (1980). Clearly the ratio M/Y is of some importance when fixing a value for U. (Where Y is the value of the recorded population under audit). If M is a large fraction of Y, U can be fixed as a small fraction of M. But if M is a very small fraction of Y, say 2% or less, U must be a high fraction of M, otherwise the audit sample size will be very large. We will return to this problem in Chapter eight.

3. Identify the total value of the recorded population (£Y)

Comment

This figure is almost invariably provided by the client.

The auditor will normally add up the individual amounts ($y_1 + y_2 \ldots\ldots\ldots y_n$) making up £Y to identify any mistakes in addition.

4. Calculate the 'skip interval'

The 'skip interval' is a concept peculiar to MUS. In order to explain this concept we must understand that MUS changes the sampling unit compared to traditional auditing. Traditional auditing selects from among a population of physical units i.e., invoices, wage slips, debtors accounts etc., while MUS selects from among a population of individual monetary units. A monetary unit is a unit of the currency used in the audit, for example a unit of pound sterling (£) or a United States dollar ($) or a German deutschmark (DM). The auditor samples a series of individual £'s but as Leslie etc (1980) puts it:

> "the auditor does not directly verify that particular £. Rather it acts as a hook and pulls the entire physical unit with it." (p. 84)

Thus the probability of selecting a particular physical audit unit is directly proportional to the value of that audit unit.

Appendix (A) provides a proof that sampling audit units with probability of selection proportional to value of the unit is equivalent to random sampling of individual monetary units.

The skip interval is calculated by dividing £U, the acceptable upper error limit at zero errors, by a factor derived from the Poisson distribution. This

	Values	Cumulative Values	Numbers Selected	Items Selected For Audit
1.	2655	2655	1375	x
2.	84	2739		
3.	107	2846		
4.	5236	8082	5475	x
5.	1723	9805	9575	x
6.	524	10329		
7.	2347	12676		
8.	186	12862		
9.	91	12953		
10,	3547	16500	13675	x
	16500			

Procedure

1. Suppose the skip interval is 4100

   we will select four items for audit since $\frac{16500}{4100} \approx 4$

2. Use random number tables to select a random start

   between 00001 and 04100. Suppose this is 01375.

3. The four selected numbers are thus:

   $$\begin{aligned} & & 1375 \\ 1375 + 4100 &= & 5475 \\ 5475 + 4100 &= & 9575 \\ 9575 + 4100 &= & 13675 \end{aligned}$$

4. If a unit value includes one of these numbers the WHOLE

   value is pulled out for audit. Thus the following items

   are pulled for audit. 1, 4, 5, 10.

Exhibit (4.8) Audit sample selection procedure in MUS sampling.

factor is determined by the level of reliability decided upon in step one above. See exhibit 4.9.

5. Add up the individual values (y_1, y_2 etc) making up the total value (Y) of the population under audit. During this process use the 'skip interval' to select the monetary unit sample to be audited.

Comment

As demonstrated in exhibit (4.8) the population under audit is added up and when the last £ of the skip interval falls into an audit unit the entire physical unit is selected for audit.

Most accounting populations are added up as a normal part of the audit. If this is the case little additional audit work is involved.

If the population would not normally be added up, additional work is involved. Various short cut techniques have been suggested by Leslie etc (1980) p.110.

Raj (1968) provides a method of PPS selection which does not require the population to be cumulated. However this two stage method may be so cumbersome for large sample sizes that cumulation may prove to be the more economic option.

At this point in the MUS sampling process the auditor will have selected a sample of n physical units. He now proceeds to audit these units using traditional methods.

6. Audit of sample so selected. List all monetary errors (e_1, e_2,..... e_n) calculate the ratio e_1/y_1 e_2/y_2 ... e_n/y_n, that is calculate the ratio of each error to the value which contains it. Prepare an ordinal ranking of these error ratios, from the largest to the smallest.

Comment

The sample selected under (5) is audited in the normal way. The ratio e/y is called the "error tainting ratio". For overstatements we assume that this ratio will not exceed 100%. Understatements could, in theory, be of infinite value. We shall discuss methods of handling this difficulty in Chapter eight. For the moment let us assume that understatements also do not exceed 100%.

7. <u>If no monetary errors are discovered the auditor can assume that the total value of error in the population under audit does not exceed £U, the upper error limit. This assumption is made with the reliability decided upon under (1) above.</u>

If no monetary errors are found nothing more need be done. The total error value is taken to be less than £U and £U was fixed as being less than materiality £M. The auditor can accept the population under audit as being of acceptable quality.

This decision is based on a simple Poisson decision process. (10)

If some monetary errors are found in the sample the auditor must adjust £U, the upper error limit, upwards to take account of the errors discovered. This must be done if he wishes to maintain his reliability, R, at the same level as when no errors were discovered.

The most difficult problem in MUS is deciding on the appropriate adjustment to make to the UEL if some errors are discovered in the sample. The errors discovered in the sample can affect the auditors prior belief about both the rate of error and the average 'tainting' of units in error.

All MUS systems must set up allocation rules for computing the adjusted UEL if monetary errors are discovered in the sample.

One method of doing this is as follows.

8. Consult a table of incremental Poisson factors, such as that shown in Exhibit (4.9). This table shows the incremental Poisson factor for 1, 2, 3..... n monetary overstatement errors discovered. Let p be the Poisson increment for error e

The additions to £U for each error discovered are calculated as follows:

Addition for first error $= e_1/y_1 \times p_1 \times Y = U_1$

" " second error $= e_2/y_2 \times p_2 \times Y = U_2$

" " ith error $= e_i/y_i \times p_i \times Y = U_i$

The original upper error limit £U is adjusted as follows.

$U_o + u_1 + u_2 \ldots\ldots u_n = U_A$

$£U_A$ is the adjusted upper error limit

If $U_A <$ £M the auditor accepts the population

If $U_A \geqslant$ £M the auditor must take further action

Comment.

The error ratios are listed in order of size from largest to smallest. Thus a large absolute amount of error might represent a small ratio.

As Garstka (1977) has written,

"the increase in the precision limit in going from the (i)th to the (i + 1) st sample error is entirely attributable to errors of size (i + i)."

This method of calculating the adjusted UEL assumes

Reliability) Confidence level)	63%	86%	95%
Errors discovered (upper bound)	Precision adjustment factor.		
0	1.00	2.00	3.00
1	1.15	1.51	1.74
2	1.12	1.38	1.56
3	1.10	1.31	1.46
4.	1.09	1.27	1.40
etc.			
(lower bound) *			
1	0.45	0.14	0.05
2	0.82	0.49	0.30
3	0.88	0.62	0.46
4	0.90	0.69	0.54
etc.			

Exhibit (4.9) Table for calculating upper/lower bound on error value.

The tables are calculated from the Poisson distribution.

* used in CMA and MEST system only.

that every undetected erroroneous £ is 100% tainted. This might be considered a conservative assumption when we consider the evidence presented in Chapter three.

Mottershead (1980) for example, has suggested that each undiscovered erroroneous £ should be assumed to be only 50% tainted.

The CMA version of MUS calculates U_i differently by ranking errors in order of absolute value and multiplying them by the skip interval (11) p. 28 - 29.

The CAV version of MUS adopts a different approach. It calculates the UEL from attribute tables for the actual number of errors discovered in the audit sample and deducts an allowance if the errors are not 100% errors (12).

Whatever method is adopted for calculating the adjusted UEL the error values discovered in themselves are unlikely to contribute a relatively large part of the UEL unless many are discovered.

All the methods of calculating the adjusted UEL studied by the author seemed to be conservative for beta risk. This factor will be discussed further in Chapter eight.

9. If understatement errors are discovered carry out exactly the same procedure as for overstatement errors. Do not net understatement against overstatement UEL's consider each separately.

Comment

The two most widely used versions of MUS, the DUS and CMA systems handle understatement errors quite differently.

The DUS system calculates an adjusted UEL for both, then deducts the most likely understatement error (13) from the UEL for overstatement and deducts the most likely overstatement error from the UEL for understatement. The logic of this approach has not, so far as we are aware, been explained (14).

The CMA system calculates a lower error limit for understatement errors and nets this off against the upper error limit for overstatement errors to arrive at a net UEL. (revised monetary precision).

Both approaches are conservative but the CMA system is more conservative than that used in DUS.

However, we must note that the reliability attached to the net bound calculated in CMA is less than the reliability attached to either of the separate bounds. Roberts (1978) p.125 states that the reliability of the net bound lies between R and (2R - 1) where R is the reliability attached to the separate bounds.

10. Study the adjusted upper bound on error value on overstatement and understatement and compare the two bounds to the materiality amount £m. Decide whether to accept the accounting population under audit or take further action.

Comment.

Ultimately all audit decisions depend on judgement based on experience. The MUS method provides a statistical framework which reduces the amount of judgement required but does not abolish it. If U_A is close to £M, either above or below, the auditor may decide to go against the decision suggested by the MUS method. If the UEL of overstatement error and understatement error are both large but of about equal magnitude the auditor is unlikely to net them off and so accept the population, he is much more likely to decide that he is dealing with a population of low quality and take further action.

The further action could be to increase sample size, to ask for the account to be rechecked, to make a provision for the most likely error, or, as a last resort, to qualify the audit certificate.

4.4. The formulae and the auditor

The level of statistical knowledge required to derive the formulae set out in this chapter is, for the most part, above the level required in professional accounting examinations.

This fact need not present an insurmountable obstacle to applying SS since a pre-programmed computer can be used to process the formulae. The auditor need only set up the parameters of the problem such as materiality and reliability. The computer can do the rest. The computer can be programmed to calculate sample size, decide the number of strata required etc., and evaluate the result of the sample test.

It is essential, however, that the auditor is taught the limitations inherent in each of the sampling methods set out in this chapter.

We shall discuss these limitations further in Chapter eight.

4.5. Conclusions.

We have studied the logic behind several forms of sampling employed in audit work.

We have noted that these methods can provide an estimate or a decision. The auditor must choose which context better suits his needs.

We have also noted the need to balance off beta against alpha risk. Sometimes this is not easy to achieve at a reasonable sample size.

With value sampling we noted a trade off between simplicity of operation and cost of accuracy. Stratification methods reduce sample size at the cost of greatly increased complexity.

Fortunately the MUS system appears to be both simple to operate and economical in sample size.

The increasing use of the computer to process accounting data will, hopefully, solve the problem of operational complexity, but the auditor must beware against using sophisticated statistical techniques without first ensuring that he understands the various limitations on their use.

the accounts under audit. Thus materiality could constitute any fraction of the value of the population under audit, it could even exceed the total value of this population!

2. Decide on an acceptable upper error limit (UEL) on the estimate of error value in the population under audit assuming that no monetary errors are discovered in the audit sample.

Comment

The UEL must be less than materiality (£M) otherwise the MUS sample becomes a discovery sample. If the sample were a discovery sample this would imply immediate rejection of the population if even one monetary error were found. This limitation can be put another way by saying that alpha risk is too high if £M = U. Kaplan (1975) provides a useful table which illustrates this point. Many writers on MUS have suggested that the value of the UEL should be fixed at 50% of the value of materiality under normal audit conditions. See Leslie etc (1980). Clearly the ratio M/Y is of some importance when fixing a value for U. (Where Y is the value of the recorded population under audit). If M is a large fraction of Y, U can be fixed as a small fraction of M. But if M is a very small fraction of Y, say 2% or less, U must be a high fraction of M, otherwise the audit sample size will be very large. We will return to this problem in Chapter eight.

3. Identify the total value of the recorded population (£Y)

Comment

This figure is almost invariably provided by the client.

The auditor will normally add up the individual amounts ($y_1 + y_2 \ldots\ldots y_n$) making up £Y to identify any mistakes in addition.

4. Calculate the 'skip interval'

The 'skip interval' is a concept peculiar to MUS. In order to explain this concept we must understand that MUS changes the sampling unit compared to traditional auditing. Traditional auditing selects from among a population of physical units i.e., invoices, wage slips, debtors accounts etc., while MUS selects from among a population of individual monetary units. A monetary unit is a unit of the currency used in the audit, for example a unit of pound sterling (£) or a United States dollar ($) or a German deutschmark (DM). The auditor samples a series of individual £'s but as Leslie etc (1980) puts it:

> "the auditor does not directly verify that particular £. Rather it acts as a hook and pulls the entire physical unit with it." (p. 84)

Thus the probability of selecting a particular physical audit unit is directly proportional to the value of that audit unit.

Appendix (A) provides a proof that sampling audit units with probability of selection proportional to value of the unit is equivalent to random sampling of individual monetary units.

The skip interval is calculated by dividing £U, the acceptable upper error limit at zero errors, by a factor derived from the Poisson distribution. This

INSTRUCTIONS FOR CALCULATING A DOUBLE SAMPLING PLAN.

1. Calculate MUER/MURR = R.
2. Find R in the first column, or next higher value.
3. Read off.

   c = acceptance number for the first sample
   k = rejection number for the first sample
   d = acceptance number for the double sample

4. The rejection number for the double sample is found by adding 1 to the acceptance number for the double sample.

5. The plan is designed so that both samples are of equal size.

To calculate this sample size we divide the figure in the last column by the MURR.

Example.

1. MUER = 3%, MURR = 1%, MUER/MURR = 3.00.
2. Find 3.0 in the first column.
3. c = 3, k = 7, d = 9.
4. The rejection number for the double sample is 9 + 1 = 10.
5. The sample size is 2.77/0.01 = 277.

Thus the final sampling plan is,

First sample 277. Accept on 3, reject on 7.
Second sample 554. Accept on 9, reject on 10.

Table A For calculating double sampling plan. (The level of confidence for both MUER and MURR is 95%.)

MURR	c_1	k_1	d	$n_1 p_1$
15.1	0	1	1	0.207
8.3	0	2	2	0.427
5.1	1	3	4	1.00
4.1	2	4	6	1.63
3.5	2	5	7	1.99
3.0	3	7	9	2.77
2.6	5	11	13	4.34
2.3	6	13	16	5.51
2.02	9	17	23	8.38
1.82	13	23	32	12.19
1.61	21	34	50	20.04
1.50	30	45	69	28.53
1.336	63	83	138	60.31

Source: By permission, from Industrial Statistics, by Paul Peach, copyright 1947, Edwards and Broughton Company, Raleigh, North Carolina.

Appendix B to Chapter four. Table for double sampling plan.
MUER= minimum unacceptable error rate.
MURR= minimum unacceptable rejection rate.

TABLE 1

Sample size n	Sample Proportion p. 0.00	.01	.02	.03
		90% Reliability		
100	.0228	.0383	.0524	.0656
200	.0114	.0264	.0396	.0520
300	.00765	.0221	.0348	.0469
		95% Reliability		
100	.0295	.0466	.0616	.0757
200	.0149	.0311	.0452	.0583
300	.00994	.0256	.0391	.0518

Appendix C to Chapter four

UEL's for various sample sizes, error proportions and reliability levels.

Binomial probabilities - one sided.

This table tells the auditor that if the proportion of the sample in error is p% then there is an r% probability that the proportion of errors in population does not exceed the proportion given in the above table.

For example if the auditor finds 1% of a sample of 200 audited items in error he can be 90% confident that the proportion of erroneous units in the population under audit does not exceed 2.64%.

Chapter 5.

A brief history of the literature and development of statistical sampling methods as applied to external auditing.

5.1. Introduction

The earliest attempt to apply scientific sampling principles to audit work appears to have been made by Lewis Carman (1933). This initial suggestion was not followed up and no further progress was made in applying SS to audit work until after the end of the second world war in 1946.

The introduction of SS techniques to external auditing has been slow and spasmodic. The development of SS appears to progress in a series of waves. One can identify four specific waves of development over the period 1948-1980. Each wave was initiated by the co-operation between a statistician and an accountant interested in the statistical approach to auditing. A listing of these associations is provided in Exhibit 5.1.

Vance and Neter pioneered acceptance sampling, Trueblood and Cyert survey sampling, Hill and Arkin the practical development of estimation and discovery sampling and two teams, Stringer and Stephan in the U.S.A. and Anderson, Leslie and Teitlebaum in Canada, pioneered the theoretical development and practical application of menetary unit sampling.

5.2. The pioneers

As noted above the first accountant to suggest the application of scientific sampling methods to auditing appears to have been L.A. Carman (1933).

Carman discusses the application of SS to the discovery

Year	Accountant	Statistician	Technique Development
48-56	Vance L. (University of California)	Neter J. (University of California) Now University of Georgia.	Acceptance Sampling
51-57	Trueblood R.M. (Touche) Davidson J.	Cyert R.M. (Carnegie-Mellon University)	Mainly Survey Sampling
57-64	Hill H.P. (Price Waterhouse)	Arkin H. (Bernard Baruch College, N.Y.U)	Estimation and Discovery sampling
61-70	Stringer K.W. (Haskins Sells) (Now DHS)	Stephan FF (Princeton University)	Monetary Unit Sampling
66-80	Leslie D. Anderson R. (Clarkson, Gordon and Co. Toronto)	Teitlebaum A.D. (McGill University)	Monetary Unit Sampling

Exhibit 5.1.

The major collaborations between accountants and statisticians in developing statistical audit sampling procedures.

of fraud (a subject rarely mentioned in the later literature). Carman attempts to relate sample size to the risk of not discovering a fraud. He suggests the following well known formula to calculate the risk.

$$R = \left(1 - \frac{n}{N}\right)^d$$

where R = probability of detecting no fraudulent units.
n = sample size.
N = population size.
d = number of fraudulent units in the population.

Vance and Neter (1956) have criticised this paper on the grounds that the discovery of fraud is not an important objective of external auditing and that if fraud is suspected a 100% check should be made. Neither criticism invalidates Carman's main contribution, namely the attempted quantification of audit sample risk.

Prytherch (1942) follows Carmen in trying to provide a statistical answer to the problem of how much test checking is required in auditing. This is the first article to discuss the application of SS to normal external audit work. The author, however, is wrong in assuming that accounting errors are normally distributed. Abrams (1947) follows up Prytherch's article by recommending the use of the Poisson distribution in audit work. He emphasises the importance of searching for all types of error (not just fraud) and recommends stratification by dollar value, "the % of items examined in each subgroup will vary with dollar value". (p.648).

From 1947 onwards papers on the subject of applying SS to audit work become more common.

Jones (1947) discusses the application of SS to test checking of clerical procedures. He suggests a form of acceptance sampling. "The acceptance level may be tentively set at the average performance of the best one third of... clerks". (p.5).

He introduces the idea of devising an operating characteristic curve suited to auditing, and emphasises the important point that the objective in auditing is to test the quality of an accounting population not to estimate an error rate (p.7). He introduces the important distinction between random and systematic error.

We should note his claim that the first application of SS to control of clerical work was initiated by W.E. Deming on verifying the punching of machine cards at the Bureau of Census (1).

Magruder (1950) describes the application of SS in reconciling the physical check of telephone equipment installed to the accounting records. A three strata sampling plan is developed. The sampling approach is claimed to achieve a sufficient degree of accuracy at only 10% of the cost of a full survey.

Rosander (1951) describes the application of SS to sampling the 1949 Income Tax returns in the United States. An interesting stratification plan is developed. The sample is drawn systematically from "flow points", there being no static population. Rosander claims that similar statistical methods had been used by Insurance Companies since 1939.

C.W. Churchman in "Railway Age" (1952) describes an application of SS to distributing inter-line charges. A 9% sample of 23,000 waybills is drawn to calculate the charges. A complete analysis cost $5000, the sample provided an answer accurate to within 1% for $1000.

We note that at this early stage in the development of SS most of the key problems that currently engage the interest of practitioners and theoreticians were already introduced. Namely, what are the most suitable statistical distributions to use, what are the most effective methods of stratification and what is the affect of the distribution and type of error on sampling design.

Before leaving the pioneers we should note the early study of accounting errors by Gregory (1952) which was earlier described in Chapter three. He studied the frequency and amount of errors in suppliers invoices to a motor car manufacturer. He noted that invoices of value exceeding $500 accounted for 18% of invoices but 78% of error. "Error rate was substantially the same as the frequency rate of each size of invoice."

No further work would be published on this crucial area of audit research for twenty six years!

5.3. <u>Acceptance sampling</u>.

In June 1947 Lawrence Vance presented a paper to the Pacific Coast Economic Association on the application of scientific sampling to external auditing. If any event can be regarded as the birth of statistical auditing this must be it. The kernal of his thought is presented in his 1949 paper published in the Journal of Accountancy.

Vance advocates the use of acceptance sampling techniques based on the "underlying mathematical work" of the statistical research group under A. Wald which had worked at Columbia University from 1941 to 1948. Vance argues that all accounting populations are really only one section of an infinite population of financial events arbitrarily cut at some point in time, therefore the binomial theorem can be used to calculate error probabilities.

It is not the actual error rate but the overall <u>quality</u> of clerical work that is important. Therefore acceptance sampling, not survey sampling, is the appropriate technique to use.

5.3.1. <u>Alpha and Beta Risk</u>.

In his 1950 book Vance introduces the concept of alpha and beta risk. He defines "alpha risk" as meaning the probability of rejecting an acceptable population and "beta risk" as meaning the probability of accepting a population with an error rate so high that it ought to be rejected. The problem of designing audit sampling plans to balance these two risks was to become a major feature of articles on SS in later years. See Laudeman (1976) for a summary.

Elliott and Rogers (1972), for example, discuss the problem in depth in regard to substantive testing of value. They argue that, in the context of auditing, it is the beta risk that is the more important risk. They suggest a method of holding the alpha risk constant and varying the beta risk.

Returning to Vance in 1950. Vance advocates a tighter control of alpha risk (5%) as against beta risk (1.0%). He defines and discusses substantive and compliance error. He states that "the only measure of accounting performance which seems amenable to statistical sampling theory is the% of errors..... this is a considerable limitation." He virtually rules out the application of SS to discovering fraud. He advocates its use in testing error rates in inventory, debtors, payroll and sales.

5.3.2. <u>The first books on applying SS to auditing</u>.

Vance published the first book on SS in 1950 and later joined with J. Neter (1956) to write their classic text on applying acceptance sampling to auditing. This book provides a detailed discussion of the construction of various types of acceptance sampling tables. It fails, however, to provide guidance on how to apply these methods in a practical context. In particular it fails to examine the peculiar nature of accounting error. The book emphasises compliance testing rather than substantive testing. No analysis of the shapes of accounting populations or error value populations is attempted. The book, though admirable as an academic text, appears to us to be rather too abstract and theoretical for the practising accountant. It would seem to have had little influence on subsequent audit practice.

Sometime during the late 1940's the U.S. Army Audit Agency and the U.S. Airforce audit group studied the application of SS to internal auditing. Little has been written on this early work but several useful sets of acceptance sampling tables were published, and some early research on SS was financed from this source. See Teitelbaum (1960).

5.4. Survey sampling - the Pittsburg Group.

In his early work Vance insists that there is no need for an external auditor to estimate an actual error rate. He need only test that the error rate is not likely to exceed some upper error limit.

This position was disputed by the Pittsburg group, based on the Carnegie Institute of Technology. The members of this group were drawn from both the academic and accounting professions. The more prominent members would appear to be R.M. Cyert, W. Cooper, R.J. Monteverde and R.M. Trueblood. The last two being accountants on the staff of the accounting firm of Touche, Niven Bailey and Smart.

The Pittsburg group advocated the use of survey sampling methods in auditing. The book by Trueblood and Cyert (1957) strongly advocates this approach and, unlike Vance and Neter, they provide voluminous sets of examples. The Pittsburg group attack acceptance sampling on the grounds that audit assurance is built up from a chain of probabilities and yes/no decision systems are too inflexible to handle this type of problem. The case is put most cogently by Monteverde (1955). He writes, "It is doubtful that most auditors would be willing to articulate specific risk and quality criteria prior to undertaking an examination of the data" (p.585). He also argues that the process average (i.e. error rate) is not known in auditing and that too many lots of acceptable quality will be rejected unless the audit sample size is very large. In other words he claims that most sampling plans incur too high an alpha risk.

The criticism of acceptance sampling by the Pittsburg group is best summed up in the following quotation from Monteverde: "The audit area.......is more judgemental and a good bit less objective....than production line operations." (p.590).

In later years Finley (1978) was to mount an attack on commonly used acceptance sampling <u>tables</u> as being too inflexible to balance alpha and beta risk. He suggests using the computer to compute a suitable sampling plan and provides a table for specific values of alpha and beta risk.

Maxim and others (1976) examine the economics of acceptance sampling audit plans and provide a formula for calculating an economically optimal sample size.

Cyert (1957) suggested using the chi-squared distribution as an economical method of estimating upper error limits. This suggestion was not followed up in the later literature.

Cyert (1957) introduced multistage variables sampling into the accounting literature for the first time. The application was not strictly an accounting one. The problem was to estimate the % of valuable metal contained in blocks of ore. Cores were sampled within car loads and used to estimate the amount of valuable metal.

5.5. The publication of suitable sampling tables.

Much of the early theoretical discussion of SS appears to have been carried on between academics. If the methods described were being applied in practice it is odd that so little was written about the many practical problems of implementing SS, such as auditing balances of zero value or auditing understatements of recorded value.

The first attempt to apply SS as a general technique by a firm of accountants seems to have been made by Price Waterhouse during the period 1954-1960. (See Healy 1964). W. Hill, a partner in P.W. and Professor Herbert Arkin, New York University, developed a set of working methods which could be generally applied to external auditing. An elementary treatise on their methodology was published by Hill, Roth and Arkin (1962). But the major contribution to this collaberation was undoubtedly the excellent set of sampling tables published in 1963 by Arkin (1963) (1974).

Prior to the publication of Arkins tables the only sets of sampling tables designed for audit work were those prepared by the U.S. Air Force (the so called stop-go sampling tables) and the Brown, Vance (1961) tables designed to assist in estimating error proportions.

Arkin's (1963) tables were easy to use, even by auditors with little knowledge of statistics. The Arkin tables allow for the skewness of the sampling distribution caused by very low error rates and his acceptance sampling tables allowed an auditor, for the first time, to select a sampling plan which balanced alpha against beta risk.

5.6. <u>Discovery sampling</u>

Arkin (1961) also suggested discovery sampling as a useful audit tool if a high alpha risk is acceptable to an auditor. The method is suited to situations where "the systematic or fairly frequent violation of internal control, or manipulation, is the situation of most consequence." (p.52). He recommended that "exploratory sampling tables" produced by the audit division of the U.S. Airforce should be used. Later he provided a set of discovery sampling tables in his own 1963 book.

At a much later date Deming (1979), one of the pioneers, was to return to the fray by suggesting the use of a discovery sampling approach using a "liklehood factor" rather than a confidence interval.

Arkin's (1963) book must be regarded as a milestone in the development of SS. It explains a relatively simple system of scientific sampling in non-mathematical language and provides an admirable set of tables. Unfortunately a second volume, which was to discuss the practical problems of implementing SS, never appeared.

It is likely that many auditors attempted to apply Arkin's system but found substantial practical obstacles in the way, obstacles which were not discussed in his textbook.

5.7. Acceptance and survey sampling - an attempt at reconciliation

Arkin's 1963 book described both acceptance and survey sampling methods. Arkin did not take sides in this debate. Nowhere in his book does he discuss the philosophy of external audit, a conceptual model of auditing, or the method by which an auditor makes up his opinion. The book is essentially an introduction to statistics for auditors rather than an introduction to statistical auditing.

In 1964 A. Charnes and others, attempted to devise a sampling technique "which combines the practical advantages of acceptance and survey sampling" (p.242). "We state the sampling objective as discovering as rapidly as possible whether particular error rates are in an acceptable range and, if not, estimating with prescribed assurance what the error rates might be." (p.242).

Charnes provides an algorithm for testing the population under audit. The system was sequential sampling procedures with pre-calculated acceptance and rejection numbers. Once sample size m is reached a confidence interval is calculated on the estimated error rate.

This method would not appear to have gained acceptance in practice. Sequential sampling, as we shall explain in Chapter (8), is difficult to apply under normal audit conditions.

5.8. Ratio and difference estimates.

It has been argued that estimation sampling of variables is uneconomical in audit work because the audit effort is focused on estimating total value rather than value of differences. The auditor is seeking differences between recorded value and audited value. By ignoring individual recorded values he is throwing away useful information.

Cyert and others (1960) used a combined ratio estimate to test the value of the U.S. Air Force motor vehicles inventory.

Arkin (1963) introduced a chapter on ratio and difference estimates and Nigra (1963) provides a practical case study using Arkin's method. Burstein (1967) advocated the use of ratio estimates rather than difference estimates since: "the scatter of the ratios for the individual accounts is apt to be relatively less than the scatter of the dollar amounts." Thus "a ratio estimate will require a smaller sample size for a given accuracy" (p.845).

Burstein also points out that "unless there is a high correlation between......value of debt and value of error in this debt, ratio estimation is not suitable."

Our study of error patterns in Chapter three must cast at least some doubt on this correlation being a high one.

Akresh (1971) also advocated the use of ratio estimates to test debtors "where there is less variation in the ratios than in the related absolute values." He emphasises that a computer is needed to do the considerable calculations involved.

Both Burstein and Akresh provide case study material. Both fail to emphasise that the rarity of errors in accounting populations may make it difficult to estimate the standard error of the ratio estimate with sufficient accuracy.

Anderson and others (1974) make much of this point in criticising the auxiliary estimator approach. They point out that these methods "are generally invalid wherever larger errors could occur in the population than were found in the sample." (p.4).

Kaplan's (1973) paper makes an important contribution to the theory of using auxiliary estimators to estimate the value of error. He points to the peculiar statistical characteristics of accounting populations. In particular he makes the interesting comment that, "It is not the (low) error rate per se that leads to statistical difficulties. It is the strong correlation between estimates of the mean and the standard deviation that causes the understatement of the standard deviation." (p.257).

5.9. Stratification.

Neter (1956) shows that intuitive stratification by auditors can be far from optimal. As we noted above Magruder (1950), Rosander (1951) and several other authors noted that accounting populations were highly skewed with regard to value. They tackled this problem by advocating various simple types of stratification.

Arkin (1963) suggests the Neyman method of allocating a fixed sample size between strata and Restall (1969) suggest a simple "equal dollars in each stratum" approach.

Goodfellow and other (1974), while mounting an attack on the DUS system, argue that mean-per-unit with carefully designed stratification can produce lower audit samples than DUS. It depends, they claim, upon the distribution of error value in the population under audit.

Since some form of stratification is essential for sample estimation of accounting values it is curious that so little has been written on the subject. The only reasonably full treatment of stratification in auditing is provided in Roberts (1978) Chapter (6).

5.10 Monetary Unit Sampling

Attribute sampling methods are adequate for compliance testing of procedures but auditors are more concerned with substantive testing of value which is almost invariably regarded as an essential step in external auditing.

We have seen above that conventional variables sampling methods, using mean per unit or auxiliary estimation, are of limited use without stratification, and that stratification is a complicated procedure requiring the use of a computer.

In the early 1960's several accounting firms began to explore the possibility of designing a statistical auditing sampling system suited to the specific needs of the external auditor. The system they sought for had to be relatively simple to operate and yet be able to handle the two major problems of audit sampling, namely the low error rate and the high degree of skewness inherent in accounting populations.

Between 1960 and 1965 one individual auditor and two major accounting firms, one in the USA and one in Canada, developed such a system. They studied the mechanics of audit sampling and designed a statistical sampling technique to fit the specific needs of the external audit process. The individual auditor was A. van Heerden in the Netherlands, the two accounting firms were Haskins and Sells

(3) in the United States and Clarkson Gordon in Canada.

The statistical audit sampling systems they devised, while not being identical to one another, are very similar in approach. The method is based on a technique which statisticians call probability proportionate to size sampling (PPS) - see Cochrane (1963) - but we will call it monetary unit sampling (MUS) since this appears to be the name currently used in the accounting literature.

The first article describing MUS was published by H. Van Heerden in 1961. This primal article was in Dutch and was not translated into English until 1979, it therefore had little impact outside the Netherlands although Van Heerden presented a short description of his method in English to the 9th International Congress of Accountants in Paris in 1967.

In 1961 Kenneth Stringer, a partner in the executive office of Haskins and Sells in New York, was asked to undertake a general review of the literature on SS. "His original efforts indicated that the available techniques and methods did not solve the problem." (Rostron 1968). However, the principle of evaluating audit risk more scientifically seemed sound. Stringer approached FF. Stephan, a professor of statistics at Princetown University, who was Chairman of the New York Applied Business Statistics Group, and asked him to assist in devising a viable system.

Stephan joined an external audit team for several weeks to find out about the objectives and mechanics of external auditing. The resulting Stringer-Stephan system was given the title cumulative monetary amount sampling (CMA). It was tested experimentally in the period 1963-1964 and had been put into widespread use in the firm by 1966. (4). The method has been subject to modification and improvement since that date but the basic system has remained unchanged to the present (1981).

In Toronto, Canada, a parallel development was taking place. Two accountants, Ron Anderson and Don Leslie, partners in the Canadian accounting firm of Clarkson Gordon, worked with Dr. A.D. Teitlebaum of McGill University to develop a scientific audit sampling method. They studied the CMA system and developed a modified version which they have called Dollar Unit Sampling, (DUS) (5). The DUS system is rather less conservative than the original CMA system and so requires a smaller sample size at any given level of accuracy. Clarkson Gordon claim to employ the DUS system on about 60% of their audits in 1980. (6).

The DUS version of MUS samples individual dollars rather than individual units as in CMA but Goodfellow and others (1974) demonstrate that the method of sampling is, in fact, identical for both systems. The DUS method introduced a system of tainted attribute cell sampling (TACS) which is less conservative than CMA sampling. The alpha risk being at its greatest when the total value of error is uniformly spread over all cells.

The DHS and CG studies represent the major pioneering work on monetary unit sampling but several other accounting firms in North America have contributed to developing the method.

A major constraint on the extended use of MUS in 1970 was the dearth of publically available literature on the subject.

5.10.1 The literature on MUS

As mentioned above an article in Dutch by A. van Heerden (1961) is the first extended discussion of the application of PPS sampling to external auditing. This article was not translated into English until 1979 so it had a very limited influence at that time. Van Heerden presented a paper on the same subject in English to the 9th International Congress of Accountants in Paris in 1967, however this paper was not published in the proceedings of the Congress.

K. Stringer, a partner in Haskins Sells, New York, presented a paper on audit sampling to the proceedings of the American Statistical Association in 1963. The paper discusses some practical aspects of applying SS to auditing, but no formal proofs are attempted. Dr. F. Stephan, of the University of Princeton, is listed as having presented a paper on statistical problems in auditing to this same AMA conference but we could find no trace of this paper although a short abstract is available.

Meikle (1972) is the first presentation of an MUS method to be made publically available. The method described is similar to the CMA method. This booklet presents a method but not a proof.

Anderson and Teitelbaum (1973) is the first public presentation of CG's DUS method. No formal proof is attempted, but a heuristic explanation of the rationale for the DUS evaluation method is provided.

Teitlebaum (1973) "presents a more formal derivation of the Poisson formulas" for evaluating DUS. He also introduces the concept of the DUS cell sampling method and the "cell" bound on MUS estimation.

The first attempt at a full mathematical exposition of an MUS system was published in 1977 by Fienberg, Neter and Leitch. The paper describes an evaluation system based on the multinominal distribution but it also formally describes several other bounds: FNL published a further paper in 1978. This paper develops their thinking along the same path. Teitlebaum, McCray and Leslie (1978) criticise the two papers by FNL and extend the DUS cell method of evaluation.

Aldersley and Teitlebaum (1979) claim to prove that "every upper error limit can be found by considering certain boundary cases" and that "DUS cell evaluation is valid in identifying the upper error limit."

Finally McCray (1980) introduces a new upper bound on the DUS system (McCray bound) which appears to be more economical than previous bounds.

There are, therefore, three bounds being used currently in MUS sampling. The Stringer bound, the Cell bound, and the McCray bound.

This brings the theory of MUS up to date (1981) but other parallel developments were taking place on the empirical testing of the various MUS systems.

5.10.2 Simulation Studies on MUS

Neter and Loebbecke (1975), included an MUS method in their simulation study of the accuracy of various SS sampling methods. Unfortunately the method they used (CAV) is not used by any of the leading practitioners of MUS. Their conclusions are, therefore, of limited practical value.

Garstka (1977) ran a computer simulation of a variety of MUS bounds. He tried out 10 different methods under a variety of error conditions. He concluded that the DUS system was very stable, but conservative.

Reneau (1978) ran a simulation of five different procedures for calculating the CAV bound suggested by Neter and Loebbecke. He suggests two of his bounds to be superior to previous bounds.

An objective evaluation of all existing bounds is being currently carried out by J.H. McCray.

5.10.3. Alpha risk

Several authors have criticised MUS methods on the grounds that the more commonly used MUS methods do not evaluate alpha risk.

For example Goodfellow etc. (1974), McRae (1974) and Kaplan (1975) raise this criticism.

If alpha risk is ignored, so it is claimed, too many acceptable populations will be rejected and re-checked. This will result in overauditing, a potentially expensive procedure.

Alpha risk is less important than beta risk but it cannot be ignored. Teitlebaum etc. (1975) answers this criticism by claiming that "it is our view that (excess alpha risk) can generally be avoided by proper sample size planning." By this he means taking a rather larger sample than required in the first place.

Kaplan (1975) tackles the problem more scientifically by providing a table to calculate alpha risk given beta risk and an acceptable error value. Unfortunately this method "requires much larger sample sizes .....it would not be unusual for sample sizes to be twice as large." (p131).

Since empirical research on error values as a proportion of total population value suggests that these values frequently exceed 0.2% (see Chapter three) it is odd that the DUS and CMA methods reject populations so infrequently.

Vanecek (1978) ran an extensive simulation to test the conservatism inherent in the DUS system (for beta risk). He concludes that all versions, are conservative for beta risk. Therefore the alpha risk potential is quite high unless specifically controlled. (p.265).

McCray (1980) provides an algoritham for calculating the maximum alpha and beta risk for any DUS evaluation. A computer programme is provided for implementing the procedure.

5.10.4. New ideas on MUS bounds

Work is currently in progress on devising algorithms for calculating more efficient error bounds for MUS.

Mottershead (1980) suggests a more economical version of DUS by assuming "that each bad £ is 50% tainted". This assumption should reduce audit sample size by about one half. The evidence from Chapter three suggests that this assumption is not an unreasonable one.

The most interesting recent work on MUS has been carried out by J.H. McCray in association with DHS. McCray has devised a new bound which he calls the McCray bound. The McCray bound operates on a ranked list of error taintings discovered in the audit sample, "to calculate an estimated upper confidence bound on the amount of total overstatements."

It is claimed that the McCray bound is "always less than or equal to the Stringer bound (in CMA). When the taintings are in the range 20% to 70% the McCray bound is significantly less than the Stringer bound." (p.1.).

McCray (1980) also introduces a finite population correction factor to allow for the variation in size of units overstated relative to the sampling interval (9).

Extensive tables are provided for rapid calculation.

The McCray bound appears to be as economical in sample size as the Stringer, cell or multinominal bound and is much easier to calculate. This last fact may recommend the bound to auditors seeking a simple and economical SS method.

5.11. The Bayesian approach

From 1948 until the mid-sixties audit sampling used classical statistical method. The MUS system devised by Stringer-Stephan was the first system to challenge this classical approach. The Stringer-Stephan method provided sampling tables with reliability factors (confidence levels) as low as 55%. Since classical statistical theory rarely uses confidence levels below 90% the new method was clearly introducing a Bayesian approach to estimation.

Stringer claimed that audit sampling contributed only a part of an auditors confidence in his audit opinion. Other factors, the quality of control, the previous years audit reports etc., also contributed towards the auditors total assurance in his opinion. The total confidence required might be as high as 95% but the confidence required in the sampling inference could be less than this.

Stringer was, in fact, introducing a Bayesian approach to audit sampling, although the Haskins Sells manual for the late 1960's does not state this explicitly. (8).
The Bayesian approach introduces the concept that all possible error rates are not equally likely to occur in accounting populations (the classical assumption). The auditor has prior knowledge as to the likely range of error rates and this knowledge can be used to modify the prior probabilities on the inference from the audit sample.

If this approach is thought to be viable the confidence level required in the inference from the audit sample can be reduced.

The first explicit discussion of applying a Bayesian approach to audit sampling was made by Kraft (1968). Kraft states that his idea is to combine subjective assurance with sample information to arrive at an overall assurance. He claims that this approach can cut audit sample size by up to 70%. In reply to the obvious criticism that "overly optimistic prior subjective probabilities can lead an auditor to accept an otherwise unacceptable situation", he claims that "in fact this does not happen". The initial disagreements in prior probabilities are diminished and finally obliterated by the weight of sample evidence." (p.53).

Kraft (1968) provides a useful set of tables for calculating audit sample size under Bayesian assumptions.

Tracy (1969) argued along much the same lines. Sorensen (1969) used a Bayesian framework to calculate an economically optimal audit sample size. Sorensen argues that "the auditor does not incorporate systematically, economic consequences into his classical statistical decision rules". (p.555). This ambitious paper provides a formula for calculating an economic audit sample size.

Vanecek (1978) develops a Bayesian version of the DUS system. This approach increased the precision of the inference from any given size of audit sample. He found the increase to be greater the smaller the sample size. By using Bayesian priors Vanecek reduced the alpha risk without increasing beta risk. However, the effectiveness of the Bayesian approach decreased as the error rate increased.

Smith (1972) doubts whether prior estimates of error rates can be objective and Corless (1972) found much diversity among the prior estimates. Felix (1976) found rather less diversity. Crosby (1979) questions the use of the beta distribution to model the auditor's prior beliefs.

Deakin and Granoff (1974) link regression analysis to Bayesian inference. The auditor could "employ the results of his regression analysis to revise his prior estimates". (p.770).

A more ambitious attempt to use the knowledge gained on

analytical review to determine audit sample size is provided by Kinney (1978). Several accounting firms have used these ideas in an informal way for many years.

The U.K. accounting firm of Thompson McLintock have built their SS system around a formal Bayesian approach. The method is described in Crabtree (1976).

As noted above the U.S. accounting firm of Deloitte, Haskin and Sells, have developed a system of analytical review called STAR which employs regression analysis. They are attempting to link the findings from the STAR analysis into statistical sampling. See Stringer (1975).

Ijiri and Leitch (1980) have suggested the use of Stein's paradox, a statistical technique, as a substitute for the Bayesian approach. This interesting suggestion uses error rates found in similar populations to improve the accuracy of estimate of the error rate in the population under audit. I.L. describe their method as an empirical Bayesian approach which reduces the notorious subjectivity of the usual Bayesian approach.

They claim, we think correctly, that the Stein approach should place the auditor in a stronger position in a court of law when defending his sampling procedures.

5.12 <u>Empirical studies of external audit sampling</u>.

Almost no empirical studies have been published on the methods by which traditional audit samples are drawn. This has made it difficult to compare the relative efficiency of traditional and statistical audit sampling. Smurthwaite (1968) is a commendable exception to this stricture. He found a "substantial degree of bias" in the traditional audit samples drawn by his firm in past years.

As noted in Chapter three a few surveys of the patterns of error discovered in audited population have now been published. These studies should allow audit sampling design to be improved in the near future, but more work is needed on this topic.

5.13. Summary and conclusion.

The early statistical work of Cochrane, Deming and Wald encouraged Vance and Neter to advocate the use of acceptance sampling as an audit tool.

The Pittsburg Group of Cyert, Trueblood and Monteverde disputed this approach on the grounds that auditing was too subjective to be reduced to a simple yes/no decision. They advocated survey sampling methods for auditing.

In the late fifties the accounting firm of Price Waterhouse together with H. Arkin, developed a useful set of sampling tables specifically designed for audit sampling. Professor Arkin broadened the range of statistical tools available to auditors in his 1963 book.

In the middle sixties several audit firms decided that while scientific sampling should be applied to auditing the sampling methods available at that time were not satisfactory. After some research and consultation with statisticians they developed and applied variants of the monetary unit sampling method. The names most closely associated with this work are R. Anderspm. D. Leslie, and A.D. Teitlebaum in Canada and K. Stringer and F. Stephan in the U.S.A.

These are the major contributers to the development of statistical sampling systems in auditing. Currently the most interesting work in progress is Kinney's attempt to link analytical review with audit sampling in a more formal way, and John McCray's attempt to design a more economical sample size within the framework of the MUS system.

Chapter five - Notes

1. We note that Professor John Neter, who has played such a leading part in the development of SS in auditing, worked at the Bureau of Census in the 1940's.

2. Professor of Statistics, Bernard Baruch College, New York University.

3. Later Deloittes, Haskins Sells.

4. Personal communication from K. Stringer. See Stringer (1963).

5. Anderson and Teitlebaum (1974) pay tribute to W.E. Deming, H. Arkin, Arthur Young and Co., and Haskins Sells for providing the impetus for their research.

6. Personal communication from D. Leslie of C.G.

8. Audit sampling: a programmed instruction course. D.H.S.

9. McCray (1969) has written that "current unpublished research results comparing the CELL and the MCCRAY bound show the minimum alpha risks are almost equal. However, when one makes modest assumptions about the size of items overstated relative to the sampling interval, the minimum alpha risks for the McCRAY bound are significantly less than the CELL bound."
Letter to TWM dated January 15th 1981.

Chapter 6

A review of several surveys of the extent of use of statistical sampling in external auditing.

6.1. Introduction

This chapter will describe the extent to which statistical sampling is being used in external auditing.

Several surveys on the extent of use of SS have been published in North America but, to our knowledge, no survey has been carried out in the United Kingdom. (1).

In this chapter we will first describe the results of a survey we carried out among firms of Chartered Accountants in England and Wales. Later we will describe the results of various surveys carried out in North America.

6.2 The England and Wales survey - large firms. The sampling frame.

We used two sampling frames for selecting firms for this survey. First we obtained from the ICAEW a frequency distribution of professional accounting firms in England and Wales having at least one partner who is a member of the ICAEW. The frequency distribution of such firms was classified by number of partners. A synopsis of this distribution is shown in exhibit (6.1.).

We stratifed this population into two groups.

1. Firms with 4 partners or less.
2. Firms with more than four partners excluding firms in the Jordan's survey (see ahead).

Those firms with four partners or less were excluded from our survey. It was felt that with the limited resources available to us it would be best to attempt to formulate a more precise measure of usage of SS by the medium and large accounting firms.

The first sampling frame was, therefore, restricted to the 1036 professional accounting firms in the ICAEW list having more than four partners. Firms on the Jordan Survey

No. of partners	No. of firms having this number of partners	% of total
1	6320	65
2/4	2331	24
5/16	958	10
Over 16	78	1
	9687	100

Exhibit 6.1.

Distribution by size, of accounting firms in the U.K. having partners who are members of the ICAEW. (Size is measured here by number of partners).

Source: ICAEW 1979

list were then excluded from this sampling frame.

We obtained a list of members of the ICAEW for 78/79. This provides a topographical listing of firms including number of partners. This list was used as the sampling frame from which an unrestricted random sample of firms having more than four partners was drawn.

The second sampling frame used by us consisted of a table of firms of accountants auditing large limited liability companies. This table is presented in the Jordan's Survey of British Quoted Industrial Companies (1979).

The Jordan's list is presented in exhibit 6.2. The list appears to cover all major accounting firms operating in the United Kingdom.

The Jordan's list was stratified into two groups. The top five firms and the rest.

6.2.1. Drawing the sample.

The sample of large firms in the Jordan's list was selected in the following way.

All of the top five firms were included in the sample for interview. An unrestricted random sample of seven additional firms were selected from among the remaining twenty three firms on the Jordan's list. The number seven was determined as the maximum number we could interview given the resources and time available for the survey.

All twelve firms in the sample of large accounting firms were approached and agreed to be interviewed. A questionnaire was drawn up (see Appendix B) and an expert on audit practice in each firm was interviewed by the author.

TABLE OF AUDITORS (1979)

	AUDITORS	No. of Cos. in Survey audited.
1.	Peat Marwick Mitchell	171
2.	Price Waterhouse	110
3.	Deloitte Haskins & Sells	110
4.	Coopers & Lybrand	66
5.	Thomson McLintock	60
6.	Thornton Baker	59
7.	Whinney Murray	47
8.	Arthur Young McClelland Moores	46
9.	Spicer Pegler	43
10.	Binder Hamlyn	39
11.	Touche Ross	37
12.	Mann Judd	35
13.	Turquands Barton Mayhew	33
14.	Dearden Farrow	26
15.	Pannell Fitzpatrick	25
16.	Josolyne Layton-Bennett	21
17.	Robson Rhodes	20
18.	Armitage & Norton	19
19.	Tansley Witt	16
20.	Chalmers Impey	14
21.	Arthur Andersen	13
22.	Stoy Hayward	12
23.	Finnie Ross Wild	9
24.	Kidsons	8
25.	Hawson	5
26.	Hays Allen	5
27.	Hill Vellacott	5
28.	Safferys	5

Exhibit (6.2) List of auditors of major U.K. quoted companies as listed in Jordans Survey (1979). Table 17.

The sample of medium sized firms was selected in the following way. A random sample of 150 page and line numbers were drawn up and applied to the ICAEW list of members. The first firm after this line having over four partners was included in the sample. (2)

A list of 150 firms was drawn up. An attempt was made to contact at least one partner in each of these firms by telephone. We succeeded in contacting at least one partner in 136 of these firms. A contact ratio of 91%.

The reason for not interviewing the remaing 14 firms was as follows.

Amalgamation	3
Refusal to supply information	3
Telephone disconnected	2
Unable to contact partner	1
Firm controlled by firm on the Jordan's list	5
	14

The non-responders were all small firms of less than seven partners. We do not believe that the characteristics of these firms were so different from the respondees as to bias the sample.

The 136 firms were contacted by telephone. A partner was asked a series of short questions on the use of SS by the firm. (See Appendix C).

The objective of the survey was to measure the extent of use of SS by medium sized accounting firms in England and Wales.

We should note that some firms who claimed to use statistical sampling did no more than use random sampling to select their audit samples. For the purposes of this study we have defined using SS to mean (1) the use of statistical method to calculate sample size. (2) the use of some random technique in drawing the sample and (3) the sample results being evaluated by some statistical technique.

6.2.2. The results of the survey.

The survey of large firms.

The twelve firms interviewed replied to the questions set as follows:

1. Do you use SS in your external accounting procedures?

		T	R*
Yes	10	4	6
No	2	1	1
	12	5	7

2. How extensive is the use of SS in your firm?

We had hoped to measure the extent of use of SS by finding the proportion of audits on which it was used and the proportion of value audited by using SS. Unfortunately the firms interviewed were unable, or perhaps unwilling, to provide this information. We had, therefore, to frame our questions rather differently by classifying firms into one or other of the following four categories.

A. SS must be used unless a good reason is put forward for not using it.

B. We prefer to use SS if possible.

C. SS is only used on suitable audits.

D. SS is available for use if the auditor decides to use it.

The use of SS under each of these categories among the ten firms using SS was as follows:

Category	No	%	T.	R.
A	2	20	2	0
B	3	30	1	1
C	3	30	0	2
D	2	20	1	3
	10	100	4	6

---

* T = top 5

R = remainder

3. Which method of SS do you use in your auditing procedures?

		T.	R.
MUS type	3	2*	1
Stratified variables	1	1	0
Attribute	7	2	5
	11	5	6

Of those firms who used SS for both compliance and substantive testing three used the monetary unit sampling (MUS) approach. One firm used all three methods but predominantly the stratified variable approach. This firm used prior knowledge of the population to decide which SS method to employ.

Those firms who used SS for compliance testing had adopted some type of attribute sampling. A given sample size was drawn and audited using random sampling methods. In two cases, if the number of errors discovered exceeded a given figure the sample was extended. If the extended sample provided a result that was still unaccpetable those firms carried out further checks which did not use statistical methods.

In this case SS is being used as a scanning device to pick out those accounting populations requiring more thorough test by traditional audit methods.

One firm used a carefully designed Bayesian approach to decide on the required confidence level.

* One firm used both MUS type and stratified variable.

4. <u>Have you plans to extend the use of SS in the future?</u>
The answers to this question were interesting. Four of the sevenfirms using SS for compliance testing claimed to be studying the possibility of extending the use of SS to substantive testing. <u>In all four cases they were considering the use of the monetary unit sampling method</u>. No other technqiue of SS was under test.

5. <u>Have you tried and rejected any other method of SS</u>?
One firm had introduced variables sampling with stratification as early as 1967, for computer stored accounting populations. They discontinued the project because of problems caused by lack of programming language compatability.
No other firm we approached had tried out and rejected an alternative type of SS.

6. <u>Do you carry out any preliminary statistical work on the accounting population to be audited to see whether it is suited to SS</u>?
Every firm we approached replied in the negative to this question.

7. <u>If SS is not applied to all audits how do you decide which audits to apply it to</u>?
The three firms using MUS type sampling applied SS to most of their larger audits. The other firm applying SS to substantive testing only applied it to audits when the data was stored on a computer which was IBM compatible.
The firms using SS for compliance testing provided somewhat vague answers to this question i.e. "The size of the audit." "We use SS if the error rate is thought to be low." "We use

SS if the data is all in one place."

These selection procedures seemed to us to be rather unsatisfactory. The decision on whether or not to use SS was usually left to the audit manager rather than the supervising partner.

8. <u>How was SS introduced into your firm</u>?

In seven of the ten cases studied the pressure to use SS came from the United States branch of the firm. In one case the pressure came from Canada. In only two cases was the move towards adopting SS initiated by a British accountant. In one of these cases the British accountant had contacted an expert on the other side of the Atlantic who supplied copious information.

It is of passing interest to note that only three British accountants appear to have introduced original ideas on the development of SS in auditing.

9. <u>How do you organise the implementation of SS in your firm</u>?

In every case but one SS was under the supervision of a technical or systems partner. The exception arose where an expert on SS was made a partner supervising another area.

Again in every case one person at managerial level had been delegated the duty of supervising the implementation of SS in auditing. This person was typically young, under thirty five, with a degree in either mathematics, engineering or statistics (and almost invariably Scottish!) In only one case did the technical expert not have a mathematical background plus an accounting qualification. In this case he had a legal background plus extensive training in SS in the United States.

In every case but one the accounting firm insisted that <u>every</u> qualified auditor should be able to use SS. Every auditor was taught the application of SS early in his training.

One firm employed a different philosophy. This firm trained a specialist group in SS and this élite group set up and supervised the application of SS in each audit, (at least in its early stages). The auditors were presented with an SS audit package. The auditors knew <u>how to use</u> this package but they did not necessarily understand how it worked.

The firm using this latter method claimed that to use SS properly the user requires a knowledge of the statistical theory behind the method. This, they claimed, could not be taught to every auditor. It would cost too much in training time. Only one of the firms in the survey made use of an expert outside the firm. This firm has only very limited connections outside the UK. They employed a University statistician to advise on statistical concepts behind SS.

All the UK firms with US connections depended heavily on their US connection in the early days. Several stated categorically that their success in getting SS off the ground had depended on their US connection in providing manuals, documentation and advice based on many years of experience in using SS in auditing in the United States.

10. <u>How do you train your staff in the use of SS?</u>

Every firm in the survey used in-house training to teach

SS to their staff. No firm used external training courses. Firms were critical of the few external training courses in SS which were available. These courses were thought to be too short and too superficial. (3).

The in-house training courses were taken in two cases by staff who specialised in lecturing but in the other eight cases experienced field staff were pulled in and were expected to take the course. In only two cases did the manager supervising SS spend time lecturing on the course (his time was thought to be too valuable for mere lecturing).

The length of time devoted to these in-house training courses in SS was as follows:

Hours of tuition	No of firms
4-5	2
10-15	5
20-30	2
50	1
	10

The firms using some variant of monetary unit sampling required 20-50 hours, the firms using SS for compliance testing only 4-15 hours.

One firm which used specialists on SS required only four hours to teach the SS package to the user, but according to our US informant, 200 hours to given in-depth training to the specialists. This in-depth training was only available in the USA and covered a good deal more than SS.

The training material which we examined consisted of some very basic statistical theory and a good deal of case study work on practical application of the chosen method.

We doubt whether an auditor using the method would understand the theory behind the method after such a short exposure to the theory unless he had previous knowledge.

11. Which books do you use as teaching aids to SS?

The firms in the survey used their own books and training manuals to teach SS. None of the standard books such as Arkin (1974), McRae (1974) Anderson (1977), Roberts (1978) or Leslie etc (1980) were used to supplement the in-house texts.

The training manuals, with one exception, were strong on practical application but weak on theoretical foundations. The firms questioned stated that they considered that the standard texts were not specific enough on application to be used as training manuals. A representative reply to the question was "these books have too much on statistics and not enough on auditing." None of the firms surveyed used the various AICPA teaching manuals on SS.

12. Having you devised a set of standard documents for recording the results of the audit using SS?

As we expected every audit firm surveyed (including the two who did not use SS) had developed a standard set of audit documentation.

The quality of those varied widely. One set, adapted from a US set, was exceedingly difficult to follow, since it contained much technical in-house jargon. Another firm had employed document layout specialists to design the forms and these were a model of brevity and clarity. The application of SS can be much simplified if some thought is given to form design. SS is difficult enough without the difficulty being compounded by complex audit records, using abstruse language.

13. <u>How do you set your confidence levels in audits using SS</u>?

The confidence level is normally set as a function of prior knowledge about the population being audited. The prior knowledge consisting of information on the previous years audit, the adequacy of the internal control system, the internal audit reports and, perhaps, the type of client. In every case we examined the confidence level was set subjectively. None of our sample used a points system to decide on the required level of confidence.

In seven of the ten cases the audit senior was required to select a level from a range of levels provided in advance. These ranged from a high of 99% to an extraordinary low of 55%. The most common level cited was 90%.

The level varied between compliance and substantive testing. The level for substantive testing was generally 5% higher than that used for compliance testing of the same population.

Only three of the sample were prepared to use a level of confidence below 80%. However, all three of these firms were experienced practitioners with considerable knowledge of statistical theory.

Although the permitted range of confidence levels was decided at a high level in advance, the actual choice was usually left to the audit senior. In one case this decision was taken by the audit manager.

Naturally the level of confidence was set high in the first instance for a new audit (say 95%) and reduced in later years (say to 80%) if the auditors confidence from other sources increased.

Although none of our sample built up their level of confidence using a points system to evaluate prior knowledge etc., one firm built up its total confidence in the audit report in this way. Each input, internal control system evaluation etc., was awarded points and the total points were required to exceed a given figure before the accounts were considered to be acceptable.

14. How do you set your precision limits (confidence intervals) in audits using SS?

The precision limit adopted appeared to depend upon the nature of the population being audited rather than on prior knowledge of the adequacy of the internal control system. In the case of compliance testing an upper precision limit of 5% error was most commonly cited. The highest upper bound cited was 8% and the lowest 3%. Note that these are upper bounds on the error rate. The actual percentage of compliance errors would normally be expected to be much lower than this figure.

Several respondents pointed out that with a confidence level of 90%, an upper precision limit much below 3% would generate large sample sizes, say over 100 units. This was not thought to be economically viable.

The upper bound on error rate is presumably related to the maximum acceptable error rate. Unfortunately none of our respondents were able to explain why any given compliance error rate was taken as being acceptable. There is no published evidence linking compliance error rate to a given <u>value</u> of substantive error.

We also asked how precision limits were set for substantive testing. This introduces the concept of materiality.

We should note that only four members of the sample used SS for substantive testing of value.

All of these firms calculated their upper error limits as a fraction of materiality. The most important determinant of materiality was profit, net of tax, but other factors also affected this calculation. Materiality was not calculated in any given case by using a simple formula, but the figures 5-10% of profit, net of tax, was mentioned by several respondees.

15. <u>Do you keep a centralised data bank of errors so that patterns of error and norms of error in different kinds of accounting population can be detected</u>?

No member of our sample did keep such a record of error. This is a pity. If a major objective of external auditing is to locate error it is difficult to see how audit practice can be improved if the nature of accounting errors are not classified and analysed.

The fact that accounting errors do fall into specific patterns was noted in chapter three.

	No.	%
Use SS in some form	16	12
Do not use SS	120	88
Number of firms interviewed	136	100
Type of SS used		%
Error rate (attribute) sampling	13	81
Error value (MUS) sampling	3	19
	16	100

Exhibit (6.3)

Results of survey of 136 medium sized accounting firms in England and Wales (>4 partners).

Usage	SEC	Other
	%	%
Often	20	6
Occasionally	19	15
Seldom	6	2
	45	23
Never	55	77
	100	100
Sample size	850	250
Response rate	46%	57%

Exhibit 6.4.

Survey of the use of SS by professional accounting firms in the United States. (from Akresh 1979).

6.2.3. Conclusions on the use of SS in external auditing by larger firms.

A few of the worlds largest accounting firms operating in the UK make extensive use of SS in external auditing.

However, once we moved away from these very large firms the use appears to be much less extensive.

Estimation of error value is, for the most part, restricted to firms having strong North American connections.

Interest in SS, particularly in regard to the MUS technique, appears to be growing rapidly among the larger firms in the UK but much of the current usage is still experimental.

6.3. The England and Wales survey - medium sized firms.

This telephone survey was less ambitious than our survey of the larger firms. The objective was simply to provide an estimate of the usage of SS by medium sized accounting firms in England and Wales. The short list of questions asked is presented in Appendix A. The answers are tabulated in Exhibit (6.3).

Exhibit 6.3. demonstrates that SS is not widely used by medium sized accounting firms in England and Wales. Only 12% of the sample used SS in some form. The sample results indicate that we can be 95% confident that the proportion of medium sized firms using SS is less than 17%.

Those medium sized firms who used SS mostly employed some form of attribute sampling. That is they used SS to estimate error rate rather than error value.

Only three firms attempted to estimate an upper bound on error value. These firms all used the MUS method.

None of the firms interviewed used SS on a majority of their audits. Their criteria for deciding to use SS were conventional i.e. size of audit or computer stored data.

Many of the partners in firms not using SS expressed an interest in SS, but it was not possible to test the depth of this interest. Several had attended introductory courses in SS and three firms reported holding internal seminars to discuss the possibility of using SS in external auditing.

We conclude that SS is not widely used by medium sized accounting firms in England and Wales. Some interest was expressed in extending its use but this interest was much less than we found among large firms.

6.4. The use of SS in the United States.

Several surveys have been carried out in the United States to determine the extent of use of statistical sampling by professional accounting firms in that country.

The most recent and also the most extensive was conducted by A. Akresh (1979) on behalf of the statistical sampling sub-committee of the AICPA.

The initial survey suggested that around 50% of US professional audit firms use SS. However when "using SS" was defined more rigorously to mean statistical evaluation, not just random sampling, 45% of the sample of firms auditing SEC registered firms, and 23% of other firms were found to use SS. The more detailed figures are set out in Table 6.4.

The author of this survey concludes that "the number of firms using SS is higher than anticipated."

A questionnaire was sent to those firms using SS asking for further information. Some of the more interesting answers were as follows:

1. When did you start using SS?

	%
Before 1970	13
1970-1973	44
1974-1979	43
	100

2. What kind of SS are you using?

	large firms	% small firms
Only attribute sampling	15	74
Attributes and variables	85	26
	100	100

3. If you use variables sampling which method have you adopted?

	No.	%
MUS/PPS	11	30
Classical variables	14	38
Both	12	32
	37	100

The firms using variables (by variables sampling we mean the sampling of value populations, this will include the MUS method, which is strictly speaking, a form of attribute sampling) sampling were the firms who had used SS for the longest period.

4. Do you use a computer to assist with SS?

	No.	%	Of those using variables sampling %
Yes	54	48	76
No	58	52	24
	112	100	100

Note the high correlation between those firms using the computer for SS and those firms using variables sampling. The operation of variables sampling is much simplified if the population is stored in a computer.

5. Does the audit firm have stated policies regarding minimum and maximum size on (a) sample sizes (b) precision limits (c) confidence levels?

	Sample size	Confidence level
No such policies	55	52
Policies set down in some cases	45	48
	100	100

Range of constraints		Most representative
(a) minimum audit sample size (units)	25/100	50
(b) minimum confidence levels	50/90%	90%
(c) maximum upper precision limits	4/12%	5%

6. Do you believe that the accounting profession should set standards or issue guidance concerning attribute sampling.

	%
Yes	57
No	33
No reply	10
	100

Guidance was particularly required concerning confidence levels and precision limits.

7. If you use variables sampling, on what % of your audits do you use it?

Of the 25 firms using classical variable sampling 16 stated that they use the method on less than 5% of their audits.

Of the 22 firms using PPS/MUS variable sampling 9 use it on less than 5% of their audits.

The evidence suggests that in the United States only the very large firms use variables sampling on a significant proportion of their audits.

J.J. Joseph (1972) carried out a survey of the use of SS by external audit firms in the US in 1972. He achieved the high response rate of 76% of questionnaires distributed.

Some of his more important findings were as follows:

1. Population size was the most important criteria determining the use of SS in auditing (p.70). The minimum viable population size was thought to be over 5000 units.

2. The minimum confidence level chosen by 87% of all offices lay between 90% and 95%.

3. The maximum upper precision level chosen by over 91% of all offices examined ranged between ± 2% and ± 5% chosen by greater than 2 to 1 margin.

4. The most important criteria for deciding on level of confidence chosen was in order of importance.

	weighting.
(a) internal control evaluation.	26
(b) audit area	20
(c) preceding years audit	10

5. Offices of smaller firms welcomed the setting of SS guidelines by the accounting profession by a 3 to 1 margin. Offices of large firms were against the proposal by 55% to 45%.

6. Few of the small firms had an in-house policy of setting confidence levels and precision limits. However 30% of large firms set confidence levels and 24% set precision limits.

   The most frequently chosen confidence level was 95% and precision limit ± 5%.

   The precision limit varied from ± 1% to ± 10%.

7. The access to standard SS computer packages (for data retrieval) was found to be as follows:

	%
large firms	93
small firms	26

8. The usage of SS by firms was found to be as follows: Between 60% and 70% of the offices of small firms had used SS at some time during the previous year. However SS was only used on 18% of all audits during that year.

   The figure for large firms was 92% to 95% with 44% of all audits using SS.

9. Attribute sampling was used in auditing by a margin of 4 to 1 over variables sampling.

A survey on the use of SS by professional accounting firms was carried out by R.E. Guinn in 1973. The survey was limited to CPA offices in four south-eastern states of the USA.

The main findings of this survey were as follows.

1. 31% of respondents used some form of SS in external auditing. The breakdown by size of firm was.

	%
National	100
Regional	67
Local	29
Total	31

2. Firms were asked if their usage of SS was increasing, decreasing or remaining the same. The results were as follows:

	%
Increasing	64
No change	35
Decreasing	1
	100

3. The non-users of SS were asked why they did not use the technique. (These were all smaller firms). The answers were as follows:

   1. Too expensive.
   2. Lack of statistical expertise.
   3. Preference for judgement sampling.

Only 5% of these non-users had actually tried SS and discontinued it, so their opinion on SS is of limited value.

4. The confidence levels used varied from 90% to 99% and the precision limits from ± 2% to ± 5%.

5. When variables (value) sampling was used the type of plans used were PPS, difference and ratio estimates.

D.M. Dennis in 1972 conducted a small survey of the use of SS by professional auditing firms in USA. The survey was restricted to the use of SS in auditing debtors.

1. Thirty of 66 respondents claimed to use SS in auditing debtors. However further investigation revealed that only 11 respondents (17%) used SS fully. That is for calculating sample size, random sampling and evaluating the results.

2. The use of SS was not found to increase as the size of the audit firm increased. (An odd finding contradicted by all other surveys).

3. Five of the eleven "full" users of SS reported "slight to significant audit time reduction as a result of using this tool".

6.5. The use of SS in Canada.

The Canadian Institute of Chartered Accountants commissioned a research study on the "Extent of Audit Testing" published in 1980.

One part of this study dealt with the use of SS by professional accounting firms for the purpose of external audit. The results of the sample survey are set out below Around 30% of the 600 practitioners approached responded to the survey.

Among the respondents 27% used SS in some form on some occasions. 37% of national firms used SS and 21% of local firms.

The type of SS used was stated to be as follows (page 27)

	Compliance testing	% Substantive testing
Attribute	47	27
MUS	33	40
Mean per unit (MPU)	4	7
Stratified MPU	3	7
Ratio or difference estimate (RDE)	3	5
Stratified RDE	-	7
Unspecified	10	7
	100	100

It was noted that materiality is seldom specified as a specific value for the total audit. However, when materiality is specified, a common rule is to set precision for the purposes of SS at one half of materiality.

6.6. The use of SS in other countries

We located articles on SS in auditing in journals etc., published in the Netherlands, Germany, Belgium, Australia, South Africa and the Philippines.

We have no doubt that SS is being used as an audit tool in those countries. However, we were unable to find any survey of the extent of use of SS in those countries.

We note that the Netherlands was a pioneer in the use of scientific sampling in auditing in the early 1960's. See Van Heerden (1961).

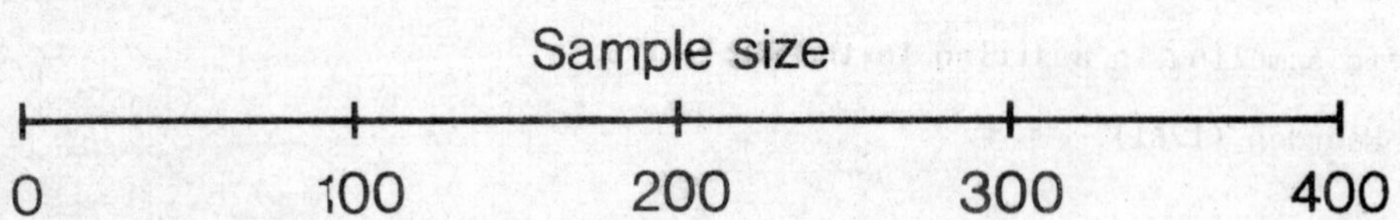

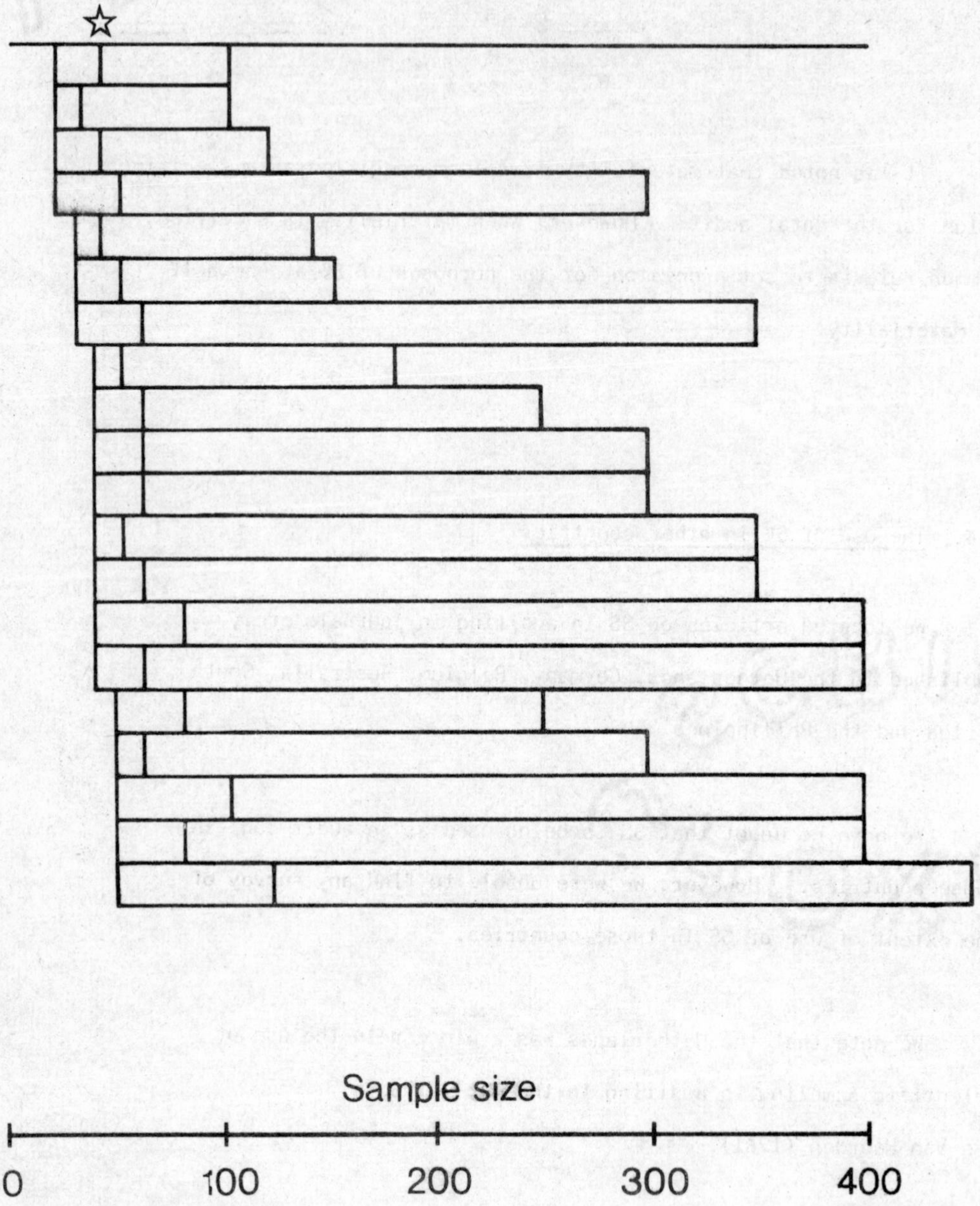

Exhibit 6.5. Range of normal audit sample sizes used by 20 accounting firms in England and Wales. The 'most common' size is also estimated. (The range is based on estimates by partners in the firms concerned.) But see note 6.

6.7. Size of audit samples.

6.7.1. The UK.

We asked twenty firms in our sample of professional accounting firms in the UK to estimate the range of size of their audit samples and the most common audit sample size used for substantive testing of audit values. (4).

The results are displayed in exhibit (6.5). Most firms stipulated a minimum sample size, usually 20 to 40 units but no maximum size.

The normal audit sample sizes ranged from 20 to 400 units but several emphasised that this was the range for normal auditing. If negligence or fraud was suspected a substantially larger sample might be tested.

The modal sample size appears to fall in the range 40 to 80. But it varies somewhat depending on the audit philosophy regarding the relative importance of sampling and systems evaluation as a testing device.

The frequency distribution of audit sample sizes appears to be skewed to the right since the modal value is closer to the lower bound.

6.7.2. North America

A sample of frequency distribution of audit sample sizes drawn by two large accounting firms in North America who use SS were made available to us. These are reproduced in exhibit 6.6.

We note that in exhibit 6.6(a) the audit sample size varied from 51 to 3185 (5) with a median size of 203 and a modal size of around 150.

The percentage of the population contained in the sample varied from 0.1% to 21% with a modal percentage of around 2%.

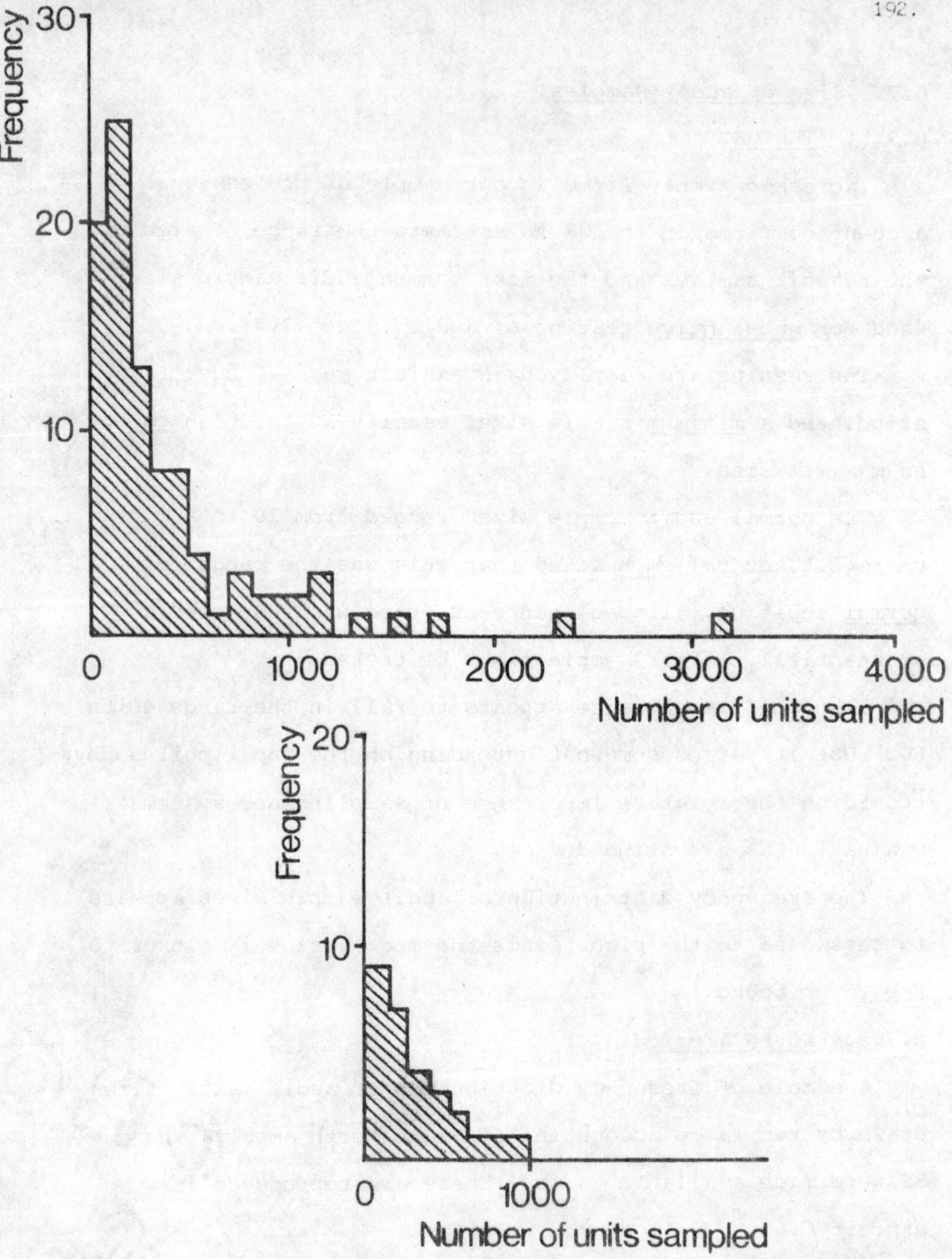

Exhibit 6.6. Frequency distribution of audit sample size, from two large accounting firms in North America. (The audits were not drawn on a random basis, but we have no reason to believe that they were not representative.)

In Exhibit 6.6 (b) the audit sample size varied from 35 to 699 with a median size of 113 and a modal size of around 90.

We note that the audit sample sizes in North America, based on this limited evidence, appear to be somewhat in excess of audit sample sizes in the UK. They are also skewed to the right.

6.8. Summary and Conclusions

This chapter presents evidence from several surveys on the use of SS in auditing by accounting firms in several countries.

The England and Wales survey suggests that SS is used by the majority of large firms. At present (1980) the usage is mainly for compliance testing but our survey suggests that a substantial expansion of SS into substantive testing is imminent.

Something of the order of 10% of medium sized firms use SS at least on some jobs, mainly for compliance testing. We did not sample the use of SS by small firms.

In North America the use of SS is more common among accounting firms. The usage rate appears to be about double that in England and Wales and the application of SS to substantive testing is more common.

The size of audit samples in North America appears to be somewhat in excess of audit sample sizes in England and Wales.

Notes on Chapter six.

1. Birmingham Polytechnic (Research in auditing group) carried out a study of audit techniques in England and Wales in 1977 which included a few questions on S.S.

2. If this firm was already in the sample the next firm of requisite size was selected.

3. The survey was carried out before the excellent three-day course designed by Arthur Young and based on the Clarkson Gordon course was published and presented.

4. These firms were a subset of the two samples previously described - 10 large, 10 medium.

5. Clearly this was not a normal sample but an investigative sample.

6. These were "off the cuff" estimates by partners. Many respondents were reluctant to provide an estimate. We do not, therefore, believe that too much reliance can be placed on these figures.

APPENDIX A.

Questions asked on telephone survey, of medium sized firms.

1. Do you use SS in your auditing procedures?
   If yes - go to question 2.
   If no - go to question 7.

2. Do you use SS for,
   3.1 Drawing a random sample.
   3.2 Calculating sample size.
   3.3 Evaluating sample results?

3. Do you use SS for,
   4.1 Estimating error rate.
   4.2 Estimating error value?

4. What method of SS do you use?
   Estimation sampling of attributes.
   Acceptance sampling.
   Mean per unit (with stratification).
   Monetary unit sampling.
   Other.

5. On roughly what proportion of audits do you use SS?

6. Why did you decide to use SS on these audits?

7. Have you ever considered using SS?

8. Do you think you will use SS in the future?

Chapter 7.

The economics of audit sampling.

7.1. Introduction.

The cost of audit sampling is an important determinant of the decision to adopt SS. SS provides the minimum average sample size to satisfy the objectives of the audit.

SS spreads the total audit sample in an optimal way over all accounts audited.

Some writers have justified the use of SS on the grounds that, on the average, it reduces sample size (1). Our own investigations suggest that many firms do not use SS because they believe that the method will result in a significant increase in sample size (2).

Chapter six illustrated the distribution of audit sample sizes in several UK and two North American audit firms. This limited evidence seems to suggest that audit sample sizes are significantly larger in North America.

Since the evidence presented in Chapter six suggests that accounting firms in the UK and North America require similar levels of reliability and materiality the smaller average size of audit samples in the UK suggests that sample size may be an important constraint on the adoption of SS in the UK. Chapter ten is devoted to a discussion of this problem.

In this Chapter we will study the economics of audit sampling, list the various operations making up the sampling process, assess their relative cost, and try to assess the sensitivity of total audit fee to a given increase in sample size.

7.2. The cost of external auditing

Exhibit 7.1. illustrates the external audit cost per

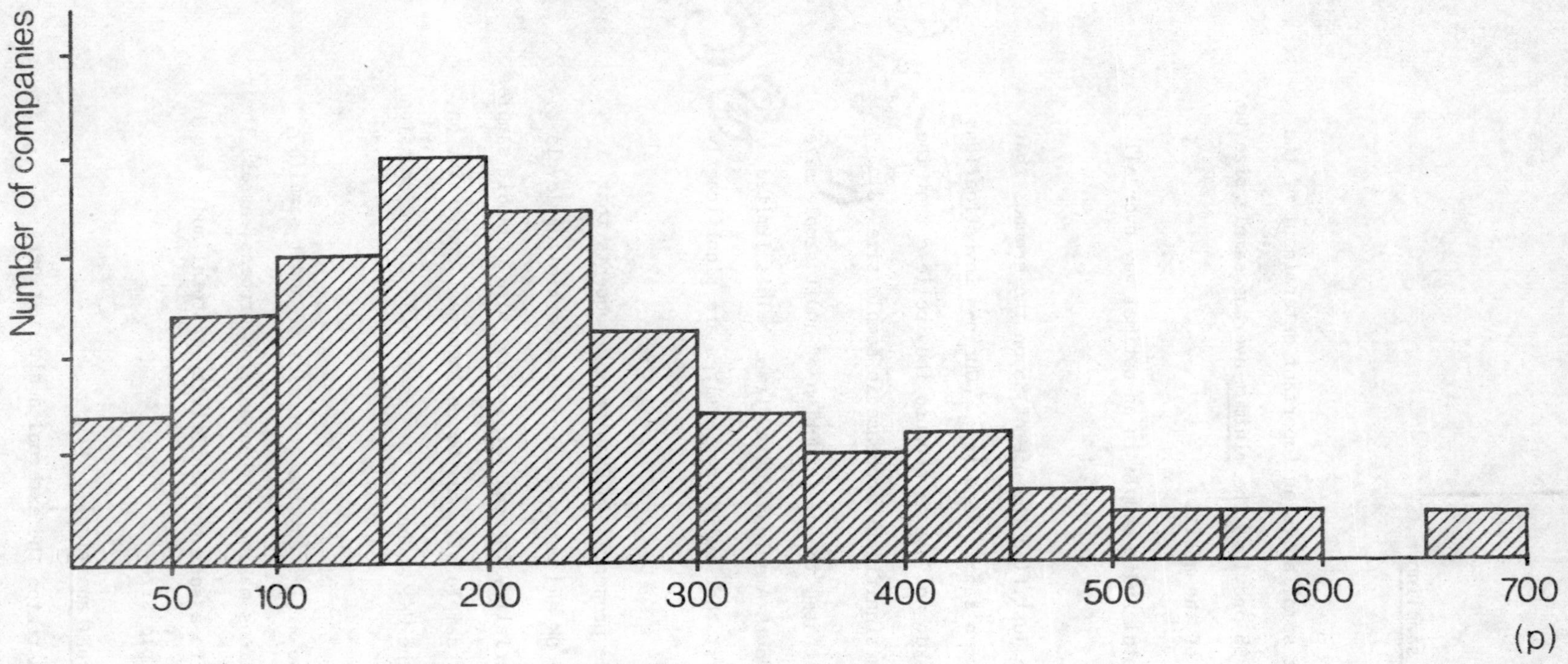

Number of companies

50 100 200 300 400 500 600 700 (p)

Exhibit 7.1. Audit fee per £1,000 of capital employed.
A study of 203 UK Limited Liability companies.
Source: Adapted from Evans (1980) p.104.

£1,000 of capital employed for a range of 203 UK companies.

We note that external audit cost varies from 8p to £9.67 per £1,000 of capital employed. The modal fee was £1.70p per £1,000 of capital employed.

Further study found a significant difference between audit cost per £ of capital employed in different industries. The study also found that the audit fee was significantly affected by the number of subsidiaries in the group (3). These findings suggest that the external audit fee is a relatively minor cost to UK companies. However this fact should not be interpreted as suggesting that a substantial increase in audit fee would be readily accepted by the client. Our discussions with audit partners pointed to precisely the opposite conclusion.

7.3. Audit sampling operations.

Audit sampling consists of the following operations,

(a) Identify the accounting population to be audited.

(b) Collect and evaluate prior information about this population.

(c) Use (b) to decide on the reliability required in the inference from the audit sample.

(d) Use the amount of materiality for this audit to decide on the upper error limit (UEL).

(e) Use (c) and (d) to calculate sample size.

(The size of sample has no affect on the cost of any of the above activities. In other words the cost of these activities are fixed relative to sample size).

(f) Generate a set of random numbers to identify units to be sampled.

(g) List the identifying numbers of the units to be sampled.

(h) Draw the sample.

(i) Audit the sample.

(j) Replace the sample.

(k) Evaluate the results of the sample test.

Activities (f) to (j) <u>are</u> affected by the size of sample. However, if a computer is being used in the audit, the cost of activities (f) and (g) are little affected by sample size. Activity (j) is often performed by the client as is increasingly activity, (h).

Thus only the cost of activity (i) is indisputably affected by an increase in audit sample size. Activity (i) is invariably the most time consuming of all the audit sample activities listed, but we emphasise that a significant fraction of the time spent on audit sampling is <u>not</u> affected by the size of the audit sample.

Our discussions with experienced auditors suggest that something of the order of one third of audit sampling time is taken up with activities other than activity (i).

We also recall from Chapter two that a great deal of audit time is spent on activities other than sampling, for example conducting the analytical review and reviewing internal control procedures.

We conclude that a given increase in audit sample size will not cause a proportionate increase in audit cost. The increase in audit cost will be <u>less than</u> proportional to the increase in audit sample size.

Audit cost job sheet

Cost head	Hours	Cost per hour £	£	%
Partner	10	100	1000	8
Manager	30	40	1200	9
Senior auditor	200	10	2000	15
Junior auditor	800	5	4000	30
Comp. operator	30	5	150	1
Specialists	10	10	100	1
Accommodation			1000	8
Travel			600	5
Other (including computer use)			500	4
			10550	81
Overhead			2500	19
			13050	100

Exhibit 7.2. A representative audit cost job sheet.

7.4. The composition of the audit fee.

Accounting firms were reluctant to provide us with detailed costings of their audit fees. However, one firm provided us with a "representative" job cost report. This is reproduced in exhibit 7.2.

Our enquiries suggest that it is the junior auditor's time that mainly is affected by an increase in audit sample size. (4). The cost of the junior auditors time makes up only 30% of the reported audit cost. Also, as we noted above, a significant fraction of the junior auditor's time, say one third, is not spent on auditing sample units.

We conclude that a 100% increase in audit sample size, if it is programmed into the audit from the start, is likely to increase the audit fee by something of the order of 25%.

If the increase in sample size is decided upon after the initial audit is complete, the economic impact could, of course, be quite different (5).

7.5. Set up cost.

Once the SS system is learned we could not identify any significant difference in set up cost between SS sampling and traditional audit sampling. In both cases standard documentation is needed, the sample results must be recorded, and the results of the test evaluated.

The SS system is more highly programmed and the evaluation more formal and more objective but these differences do not seem to generate additional costs.

The cost of computer use does not seem to be significant in most cases. (6). We were quoted a few pound sterling per audit.(9) This conclusion assumes that tape handling is run in tandem with a normal run by the client.

7.6. Learning cost

If an accounting firm decides to adopt SS in it's audit practise a learning cost will be involved.

Presumably the firm will "buy in" an experienced practitioner to introduce a well tried system.

Operational competence in SS can be taught in a few days. Compliance testing systems can be taught in one day, monetary testing systems require two to three days. (7).

We doubt whether the set up or learning costs, by themselves, would be likely to be the determining factor in deciding whether or not to adopt SS, even in a small firm.

We interviewed a partner in several small accounting firms in the United States (less than five partners) who claimed to use SS in external auditing on a regular basis.

7.7. Summary and conclusions

The decision to adopt or not to adopt SS is likely to be strongly influenced by prior beliefs about the cost of SS compared to the cost of traditional audit sampling.

The cost of SS was frequently mentioned in our discussions with partners of medium sized accounting firms.

The reluctance of professional accounting firms to provide us with detailed costings of specific audits hampered our ability to research this topic. However, using evidence, personal experience and the limited data available we came to the following conclusions.

1. The audit fee is a minor cost to limited liability

companies in the UK. The modal fee being £1.70 per £1000 of capital employed.

2. The cost of many operations in the audit sampling process are not affected by the size of the audit sample.

3. An increase in the audit sample size is likely to result in a less than proportionate increase in audit cost. We calculate that a 100% increase in audit sample size, so long as it is programmed into the audit from the start, would increase total audit cost (and so audit fee?) by about 25%.

4. The set up and learning costs of SS are economically significant but we doubt whether these factors, by themselves, would decisively influence the decision to use SS. (We know of several very small firms in the USA, with less than five partners, who use SS on a regular basis for external auditing).

Once SS is set up it is no more expensive to operate than traditional sampling so long as the size of the audit sample is not increased.

The limited evidence available (see chapter six) suggests that audit sample sizes in North America are larger, on the average, than in the UK. Therefore if reliability and materiality are similar in both countries sample size may be a constraining factor on the extended use of SS in the UK (8).

This chapter has argued that an increase in audit sample size in the UK would result in a less than proportionate increase in audit fee.

We will provide alternative solutions to the problem of excess audit sample size in chapter ten.

Notes.

1. See for example Aly and Duboff (1971).
2. This comment was frequently heard during discussions with partners while carrying out the telephone survey described in chapter six.
3. See Evans (1980) for a fuller discussion of this study.
4. If the audit is extended by increased sampling the accommodation and possibly travel expenses would also be increased but on the evidence available these costs would make only a small impact on total audit cost.
5. For example the accounts may be distributed over several sites far distant from one another.
6. Except in the case of stratifying a value population into a large number of strata. Here a double run is required. This can be expensive.
7. The well known Clarkson Gordon course on DUS is a three day course (18 hours).
8. It could be argued that the lower audit sample sizes in the UK are consequent on prior beliefs about the quality of UK accounting systems and staff compared to those in North America. We find this proposition difficult to sustain. We know of no evidence to support it.
9. The DHS STAR system costs £15 a run on a microcomputer and around £50 a run using timesharing. DHS Auditing and Accounting Newsletter 22.12.80.

Chapter 8

A critical study of the various statistical sampling methods used in external auditing.

8.1. Introduction

In this chapter we will discuss the applicability to external auditing of each of the statistical sampling methods described in Chapter 4.

We will compare each method against the criteria for audit sample efficiency set out in Chapter 2. We will also attempt to measure where appropriate, how closely the statistical sampling method follows traditional audit sampling methodology.

We will first examine those scientific sampling methods suited to compliance testing of procedures. Later we will examine the various methods which can be applied to sampling monetary values.

8.2. Sampling to estimate compliance error.

As explained in Chapter 4, auditors who apply SS to compliance testing must adopt one of two possible approaches.

One approach is to estimate the actual error rate in the population, this approach is called survey sampling.

The alternative approach is to test whether the actual error rate is less than some acceptable rate. This decision orientated approach is called acceptance sampling.

Since we favour the acceptance sampling approach to compliance testing of procedures by auditors, we will begin by examining this method.

8.3. Acceptance Sampling.

Acceptance sampling can be used to test the compliance of items in the accounting population with given accounting procedures. The method attempts to set limits on procedural error, not value error.

Acceptance sampling is a form of hypothesis testing. Given the four parameters, acceptable error rate, unacceptable error rate, alpha risk, and beta risk, a suitable sampling plan can be calculated from the binomial or some other suitable distribution.

The sampling plan is defined as a sample size, n, and an acceptable number of errors, c. If a single sampling plan is used then, if the number of errors discovered exceeds c, the population under audit is rejected immediately c + 1 errors are discovered. If the number of errors discovered is less than or equal to c, the population is considered to be of acceptable quality.

This method of analysing the audit sample is very similar to the method employed in traditional audit sampling where the auditor "rejects" the population on finding an intuitively derived critical number of errors.

8.3.1. Advantages of acceptance sampling

The advantages provided by acceptance sampling over competitive methods are as follows:

1. The theory of acceptance sampling is highly developed and easily accessible in many standard textbooks.
2. An operating characteristic curve for any particular sampling plan can be constructed and studied by the auditor.
3. The acceptance sampling plan provides the mini mum information sufficient to satisfy the needs of the auditor. A specific estimate of the actual error rate is not provided, only an estimate that the rate does not exceed some given figure. Since excess information costs money in the form of additional sample units, acceptance sampling, at any given level of accuracy, will

tend to be cheaper than sample survey methods which attempt to measure the actual error rate in the population under audit. We will argue later that an estimate of an actual error rate serves no useful purpose unless it exceeds the acceptable rate.

4. Acceptance sampling allows an auditor to balance alpha against beta risk. No other attribute sampling method allows the auditor so much scope in balancing these risks.

5. Once c + 1 errors have been discovered the population can be rejected <u>immediately</u>. Thus accounting populations with error rates far in excess the AER will be quickly identified and rejected. Audit time will be saved since the full sample need not be audited.

6. A procedural audit based on acceptance sampling can be programmed in advance of the audit to carefully designed <u>consistent</u> standards. Under these conditions the <u>quality of audit staff</u> need not be so high as would be required if more discretion were to be left in their hands. Thus the audit wage bill can be reduced.

7 The audit sample sizes generated by acceptance sampling using reasonable parameters are of the same order of size as traditional sample sizes used in UK audits. For example the parameters

AER 2%	Alpha risk 20%
UER 7%	Beta risk 5%

Population size: very large >10,000 units generates an acceptance sample of the order of 100.

8. Several technical advantages will be discussed under "technical points" later in this section.

8.3.2. Critique of acceptance sampling

The arguments in favour of acceptance sampling are so strong that it is not surprising that the method was the first to be suggested for use in audit work. Despite this the method has not been widely adopted by external auditors (see Chapter 6). The arguments raised against acceptance sampling are as follows:

1. Acceptance sampling does not measure value, only rate of error. No algorithm is currently available which can convert an error rate into an error value

Comment

This fact is not in dispute, but the advocates of acceptance sampling recommend that the method should be used for testing compliance with procedures, not for estimating error value.

2. Acceptance sampling does not attempt to estimate the actual error rate in the population under audit.

Comment

This is true, but those such as Monteverde (1955) who have criticised acceptance sampling on this ground have failed to explain why an estimate of the actual error rate is useful. So long as the rate is likely to be below the UER what purpose is served in estimating it? The population under audit is acceptable whatever the actual rate might be so long as it falls below the UER. The audit decision is not affected by this improved precision which costs additional sample units to obtain.

3. Acceptance sampling was designed to control the quality of small continuous lots of manufactured products where defective parts were replaced and an a priori process average of defectives was agreed. Audit populations are quite different. The audit process

is more subjective, the populations are larger and examined less frequently, error is less easily defined and identified, and a process average is not agreed. Arkin (1963) and (1974) is the best known exponent of this point of view.

Comment

There can be no doubt that industrial quality control is a very different control process compared to audit sampling, but this fact, in itself is of no importance so long as the auditor is aware of the various differences. The question is not whether external auditing resembles industrial quality control but whether the statistical process of acceptance sampling can be successfully applied to external auditing.

The various differences such as lot size, frequency of test, identification of defective items etc., are, except in one instance, of little consequence to the auditor. The one significant difference, related to the opinion of the tester, will be examined in the following section.

One specific criticism by Monteverde (1955) would seem to be invalid. Monteverde claims that acceptance sampling cannot be used because there is no process average or generally accepted AOQL in auditing. Neter (1952) points out that the average estimated error rate from previous years samples can be used as a process average and, as a later section of this Chapter will argue, deciding on an acceptable process average is a key decision in all forms of auditing.

The attempted comparison between industrial quality control and audit sampling would seem to be something of a red herring, although the exercise may provide some benefit by warning auditors against applying IQC techniques to auditing without first thinking hard about the very different context in which IQC operates.

4. <u>The decision rule derived from an acceptance sample is a binary yes/no decision rule. This is too simple a decision rule for external auditing</u>. The auditors judgement extends over a wide spectrum of possible decision, it is a continuous variable not a binary two state decision process. The information from a procedural test feeds into a complex decision process by which an auditor decides whether or not to accept a set of accounts. An estimated error rate, which is a continuous variable, is better suited to shading an auditor's opinion rather than a simple accept/ reject decision. See Monteverde (1955) or Cyert (1957)

Comment

At normal levels of assurance an error estimate with a confidence interval of $\pm$ 2% or less would be likely to generate an audit sample size far in excess of the 40-150 units traditionally drawn in the UK. If the confidence interval exceeds this figure it is difficult to see what gain in accuracy survey sampling is achieving over acceptance sampling.

Acceptance sampling can be adapted to provide a triple decision rule by using double sampling or any number of decision criteria by using multiple sampling plans. Charnes, etc., (1964) presents a sequential sampling plan which provides a sequential accept/reject decision rule up to a fixed sample size, m, and then calculates an estimate of the error rate.

With such possibilities available it is difficult to see why the other considerable advantages of acceptance sampling should be sacrificed to obtain a variables estimate rather than a simple decision rule.

We suspect that audit sample sizes in the UK will have to be increased substantially (1) if auditors wish to make reasonably accurate measures ($<2\%$) of actual error rates at normal levels of assurance.

As Finley (1978) points out, "If p, (the error rate) is slightly less than UER, the auditor will, more often than not, wind up rejecting the population". If the parameters of the acceptance sample are properly specified they encapsulate the auditors judgement while survey sampling leaves the auditor's judgement much less well defined.

5. Under real life audit conditions the UER will be much too close to the AER. This will generate audit samples which are much too large to be economic. See McRae (1974)

Comment

The closer the UER approaches the AER the larger becomes the sample size since discrimination between acceptable and unacceptable batches becomes more difficult. At the limit UER = AER and the entire population must be tested.

The auditing profession has set down no hard and fast rules regarding AER and UER. In practice AER appears to vary between 2% and 5% and UER between 5% and 10% (See Chapter 6). Obviously an auditor could choose rates so close as to make acceptance sampling uneconomic. But auditing, like any other monitoring activity, must balance the risk of error against cost. A study of acceptance sampling tables suggests that at 90% assurance an alpha risk of 20% and beta risk of 5%, the gap between UER and AER must be at least 5% to generate audit sample sizes small enough to be acceptable in the UK (and 4% for the USA and Canada).

Our survey of external audit practice in Chapter 6 suggests that such UER's and AER's are not out of line with current practice in those countries.

6. Acceptance sampling tables suited to external auditing are not readily available. In particular it is difficult to find tables from which one can select a plan which balances alpha and beta risk (Finley 1978)

Comment

This criticism was valid before the publication of Arkin's tables in 1963 but these tables are specifically designed to allow an auditor to choose a sampling plan satisfying the required alpha and beta risk.

Finley (1978) complains that Arkin's tables "are tabled in large jumps for sample sizes, reliance on the tables will often force the use of a much larger sample than is actually needed" (p.34). Also "one is sometimes unable to find any satisfactory plan".

Finley suggest that the auditor develops "a time sharing interactive computer programme" for providing suitable plans. He claims the incremental computer cost to be "under one dollar" per plan.

It would surely be reasonable to ask the Accounting Institute to commission a set of acceptance sampling tables suited to external auditing.

8.3.3. Technical aspects of acceptance sampling

The most useful attribute of acceptance sampling is that it allows an auditor to balance out alpha and beta risk.

In auditing, beta risk, the risk that an auditor may accept a population he should reject, is the more important risk. If the auditor gets this wrong, at best he will undersample on substantive testing, therefore increasing the risk factor in this part of the account. At worst he will miss a major error and find himself sued in a Court of Law for negligence. Damages for audit negligence, particularly in the USA, are now substantial. Most US firms are covered for insurance to an extent far exceeding annual audit fees.(2).

Beta risk should lie between 20% and 1% depending upon prior expectations as to the quality of internal control and the cost of making a wrong decision.

Alpha risk is less important in the sense that if a wrong decision is taken the result is simply overauditing, or possibly annoying one's client by insisting on him rechecking an acceptable account.

The acceptable level of alpha risk depends upon the cost of rechecking the rejected account. If this cost is high, alpha risk should be reduced. A degree of alpha risk lying between 30% and 5% seems a reasonable range from which to choose.

In theory an economically optimal alpha risk would be calculated by balancing the reduced primary audit cost against the increasing probability of secondary audit cost.

What is an acceptable error rate (AER) in auditing? We discussed this problem in Chapter 3 and concluded that it varied between accounts and clients.

Acceptance sampling procedures have been criticised for not providing a clear decision rule when p, the true population error rate, is predicted by the audit sample to fall between the AER and the UER. But this lack of clarity is reflected in the imprecision of audit standards themselves. There is no generally accepted minimum unacceptable error rate for different accounts and different clients. The boundary between the AER and UER is blurred. Acceptance sampling design fits this situation well.

Most of the above discussion has been directed towards single stage acceptance sampling. Both double, multiple stage and sequential sampling have been advocated for audit work.

It can be proved that average sample size required declines, at a given degree of accuracy, as the number of sample stages increases.

Sequential sampling requires the smallest average sample size.

(Vance & Neter 1956). Unfortunately the methods by which samples are drawn in auditing makes multiple and sequential sampling technically difficult and expensive. One solution would be to draw a full sample as for single sampling but to audit as for sequential sampling. Stopping the audit once the accept or reject number is reached. There would be no savings in drawing and replacing the sample but possibly considerable savings in audit time.

Finally we must mention what statisticians call average outgoing quality limit (AOQL)

If every batch which is rejected by acceptance sampling is tested 100% the defectives remaining in the system can be limited to a figure not exceeding p% (Arkin 1974). This is a powerful control device in industrial quality control and has been used in auditing by A.C. Dekkers in the Netherlands (3).

The applicability of the method is, however, severely limited in the context of external auditing. It is often not economically feasible to check every item in a rejected audit population. Also the purpose of compliance testing is usually to judge the quality of internal control, not to maximise the detection of errors (4).

AOQL is a useful control device in internal audit where batches of documents are regularly tested for compliance with given accounting procedures

8.4 OTHER PROCEDURES FOR CONTROLLING BETA RISK

If the auditor is to use SS for testing procedures he must control beta risk. Two methods other than classical acceptance sampling have been proposed for controlling beta risk. These are:

1. The "upper limit test" (ULT) suggested in SAS 320B by the AICPA (1974) p.83. (See Finley P.32)
2. The "confidence interval approach" (CIA) suggested by H. Arkin (1976). This follows the Pittsburgh Group method of survey sampling.

Both these methods may provide lower sample sizes than acceptance sampling for any given level of accuracy but unfortunately they are defective in certain respects compared to acceptance sampling.

8.4.1. The Upper Limit Test

Under this system the auditor specifies:

1. Unacceptable error rate (UER)
2. Required confidence level in the estimate
3. Anticipated error rate (AER)

The sample is chosen so that the probability of rejecting the population when p equals the UER is equal to the confidence level. (See Finley 1978) p.32.

This system suffers from two serious defects. First the alpha risk is not controlled, resulting in many acceptable populations being rejected. Second, the amount of variation in the estimate depends upon the size of the sample. But this, the sample size, has yet to be determined! A probability distribution ought to be attached to the sample estimate but it is not. Therefore the risk is not known.

This method is inadequate compared to acceptance sampling.

8.4.2. The confidence interval approach

Under this system the auditor specifies:

1. The required precision of the estimate
2. The required confidence level
3. A maximum acceptable error rate (AER)

The decision rule is as follows:

If upper bound on estimate $\leq$ AER - accept

If lower bound on estimate $\geq$ AER - reject

If AER within confidence interval - take appropriate action.

Comment

This is really a form of survey sampling adapted for the purposes of audit. It enjoys the following advantages.

1. It provides an estimate of the actual error rate: we noted above that several writers have suggested that this information is required in forming an audit opinion derived from information from several sources.
2. It provides a limited form of control on both alpha and beta risk
3. It provides a three state decision rule allowing the auditor to take further action if the result of the audit test is doubtful. (Some might think this a disadvantage)

The possible disadvantages of the system are as follows:

1. The method forces the auditor to choose alpha and beta risks which are approximately equal. This arises because the two risks must be less than or equal to one half of (1 - confidence level). See Arkin (1976). We believe, as noted above, that a strong case can be made in favour of a low beta risk being allied to a high alpha risk.
2. The indeterminate case will hand over more discretion to the auditor. If the audit staff is not of high quality this may cause some difficulty. On the other hand this delegation to a decision rule of clear cut cases and subsequent concentration on dubious cases will focus audit judgement on those areas where it is most needed.

Arkin's confidence interval approach cannot be rejected. It enjoys certain advantages over acceptance sampling and suffers from certain limitations. Each individual auditor must decide for himself whether the availability of an error estimate with a rather wide confidence interval (normally $\geqslant$ 3%) adequately compensates for his inability to vary the alpha and beta risk.

On balance we prefer acceptance sampling since we believe the CIA confidence interval to be too wide to be useful.

8.5. ACCEPTANCE VERSUS SURVEY SAMPLING

Compliance testing of accounting records may use either acceptance or survey sampling. Acceptance sampling draws a random sample and counts the number of errors, if the number exceeds c, the acceptance number, the population is rejected. The quality of internal control is considered inadequate. This will usually lead to the auditor requiring a high level of reliability for his later substantive test.

Survey sampling also draws a random sample but the auditor uses the proportion of errors in the sample to estimate the proportion of errors in the population under audit.

Acceptance sampling results in a decision to accept or reject.

Survey sampling provides information but not a decision unless Arkin's confidence interval approach is used. Even if Arkin's method is used a good number of samples are likely to suggest the indeterminate decision.

The key questions in deciding between acceptance and survey sampling are follows:

1. How highly programmed does the supervising partner wish the accept/reject decision to be? In other words, how much discretion is to be left to the staff auditor?

2. What sample size is economically feasible? If the sample size is usually in the range of 40-120, as in the UK, then estimation sampling confidence intervals are likely to be too wide to be of much use. At error rates below 10% they are likely to be of the order of $\pm$ 3%. These are surely too wide to be of any practicable use in focussing an auditors judgement. Acceptance sampling provides no less information within a more acceptable framework.

We advocate that the accounting profession should concentrate on improving acceptance sampling procedures for compliance testing (5).

Criteria		Type of Sampling		
		Acceptance	Survey	Traditional
Representative		Y	Y	N
Corrective		N	N	Y
Preventive		Y	Y	N
Protective		N	N	Y
Consistent		Y	Y	N
Simple to operate		Y	N	Y
Economic	Weak	Y	N	Y
	Strong	Y	Y	N

Exhibit 8.1.

Criteria for deciding on the relative efficiency of three audit sampling methods - compliance testing.

8.6. THE EFFICIENCY OF AUDIT SAMPLES USED FOR COMPLIANCE TESTING

In Chapter 2 we introduced a set of criteria for assessing the efficiency of an audit sample. Exhibit 8.1 measures acceptance sampling, survey sampling and traditional sampling against those criteria.

Acceptance sampling evokes six positive responses, survey sampling four and traditional sampling four.

An acceptance sample, as a random sample, has a good chance of being both representative and preventive. It is also consistent across audits and can be shown to satisfy the required audit sample conditions at minimum size, and within audit sizes used in traditional sampling. It therefore satisfies both the strong and weak economic criteria. Since the acceptance sample is a random sample drawn from a population which is highly skew with regard to value it is unlikely to be protective and the proportion of errors in the sample is likely to approximate the proportion in the population as a whole unless some form of stratification is used. An acceptance sample is not, therefore, corrective. The method is, however, simple to operate and to delegate.

Survey sampling evokes three positive responses as being representative, preventive and consistent. The reasons are the same as those noted above for acceptance sampling.

Survey sampling also follows acceptance sampling in being neither corrective nor protective. The two criteria differentiating survey from acceptance sampling are the economic and operational criteria. Survey samples providing sufficiently close confidence limits, (say $<$ 2%) tend to be much larger than traditional audit samples drawn in the UK, so survey samples fail the weak economic criteria although they pass the strong one.

Survey sampling is more complicated in operation compared to decision type sampling methods. More judgement and skill is called for from the audit clerk.

Traditional audit samples evoke only four positive responses. They are often designed to maximise the probability of detecting error. They thus satisfy the corrective criteria, but this means that by design they are not representative. Since traditional samples are not random they are neither representative, preventive nor are they likely to be consistent. By testing a higher proportion of high value items they are, however, protective. They are economic in the weak sense of being sufficiently small for the cost of sampling to come within the limits of traditional audit fees. However, there is no way of proving that they are economic in the strong sense of satisfying audit requirements at minimal sampling cost. No prior audit criteria are set up by which to judge the economic efficiency of the sample.

Traditional audit sampling is easy to operate since few specific rules are applied, detailed procedures of testing being often left to the discretion of the audit senior.

8.7. DISCOVERY (EXPLORATORY) SAMPLING

Discovery sampling is a form of acceptance sampling where, c, the acceptable number of errors, is always zero. The method was first applied to auditing by Arkin (1961).

The auditor decides on a suitable unacceptable error rate (UER) and confidence level, Tables have been made available for finding sample size. See McRae (1974) (p.239).

This method appears at first sight to provide an exceedingly economical method of testing for low error rates between say 2% and 0.5%.

Population size: Large (over 2000)

UER 3%

AER 0.5%

Sample size	Beta Risk %	Alpha Risk %
75	10	32
90	6	37
100	4	40
125	2	48
150	1	55

(No discovery sampling plan combines a beta risk of 10% or less with an alpha risk of 20% or less)

Exhibit (8.2) - Discovery sampling. Impact of audit sample size on alpha and beta risk.

Population size - Large (at least 2000)

UER 3%

AER 0.5%

We suggest that an alpha risk exceeding 20% would be unacceptable to an auditor.

Unfortunately this initial promise quickly evaporates when subjected to the bright light of statistical analysis.

8.7.1. Comme ntary - Apparent Advantages

Since the sample sizes required at any given level of discovery sampling (DS) are relatively low and the method of accepting the population, if no errors are found, follows closely on traditional audit method, discovery sampling seems to provide an almost ideal sampling method for the external auditor. DS also provides the useful benefit that sampling stops once one error is discovered.

8.7.2. Critique

Unfortunately DS suffers from one severe defect. The method entirely ignores alpha risk. It can be demonstrated from acceptance sampling tables (see Arkin 1974 - Table F) that with a UER as high as 3% and a beta risk as high as 10%, the alpha risk at an AER of 0.5% lies between 30% and 50% If the confidence level is raised or the AER increased the alpha risk is substantially increased - see Exhibit 8.2.

Thus the risk of rejecting 'good' populations is much too high unless the actual error rate is very low indeed, say 0.1%.

The economy in sample size is therefore bought at a very high risk of over auditing.

We conclude that discovery sampling is only suited to those audit situations where some populations are anticipated to have virtually no procedural errors and the auditor wishes to provide support for this view. It can also be used to differentiate audit populations into two groups when the variability of error is large between the groups.

We doubt that discovery sampling can be applied to normal external auditing except perhaps to those institutions with very high standards of internal control such as banks.

In favour of DS it should be noted that the excessive risk attached to discovery sampling is the risk of overauditing. The lesser of the two sampling risks.

8.8 SAMPLING TO ESTIMATE MONETARY ERROR

The sampling procedures discussed to date are used, for the most part (6), to test compliance with accounting procedures.

The main objective of the external audit of accounts, however, is not to test for quality of internal control but to form an opinion as to whether the accounts give a "true and fair" view of the financial state of the Company under audit. This objective can only be realised if the auditor is of the opinion that the value of error remaining in the population after the audit is completed is not material.

Thus an external auditor must test the accounts under audit to assess at the very least, an upper limit on value error.

Since many accounting populations are made up of a large number of value transactions the auditor must use some form of value sampling.

The sampling of value populations, where each unit can take on any one of a wide range of values is more difficult, both theoretically and practically, than is sampling of procedures. Procedures can have only one of two values attached to them. They are either right or wrong.

As described in Chapter 4, the following methods have been used to sample accounting populations and estimate the degree of value error in these populations.

(a) Mean per unit estimation

(b) Difference estimation

(c) Ratio estimation

(d) Monetary unit sampling

Stratification procedures have been applied to improve the sampling efficency of (a), (b) and (c).

Regression estimates have also been used to estimate error value.

The remainder of this chapter is devoted to a critical examination of those various sampling techniques.

8.9. MEAN-PER-UNIT ESTIMATION

Since the main objective of external auditing is to test the accuracy of the total value of an item recorded in the accounts, it might seem that the simple and well known method called mean-per-unit estimation would be the best sampling method to use.

Two approaches are available when using the mean-per-unit method. The first approach is to estimate the total value of the audited amount The second approach, a decision style approach, is to judge the reasonableness of the recorded amount.

Both approaches have been used in practice.

When usi ng the mean-per-unit approach the auditor must decide, prior to sampling, on the confidence level and precision required of his estimate. The standard deviation of the audited population needs to be estimated and from this information the size of audit sample is calculated.

The theory behind the method is well established and easily accessible in standard textbooks. See for example, Cochrane (1963).

The estimate of the audited value can be measured to any required degree of accuracy, but at increasing sample cost.

8.9.1. Advantages of the mean-per-unit method

The mean-per-unit method is simple to operate and well tried outside auditing. Also it can be used when the individual recorded values making up a total account (a) are not available or (b) are difficult to audit or (c) show a very low correlation, say less than 0.5, with the audited amounts. Such conditions are unusual.

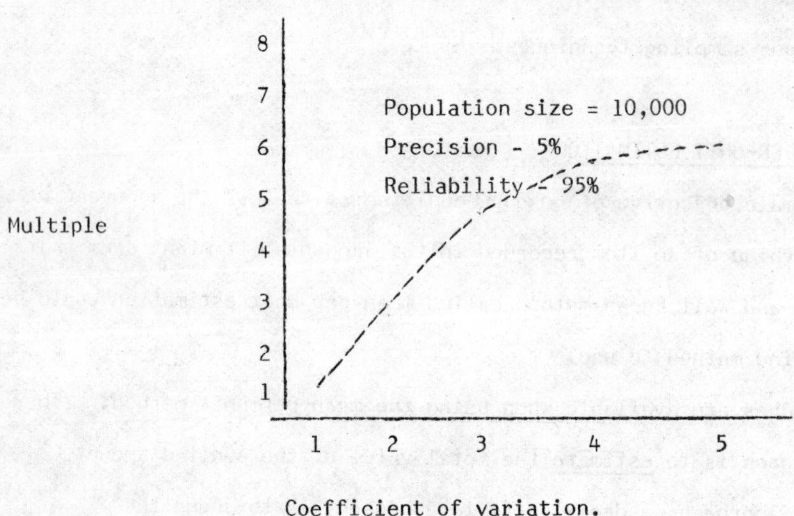

Exhibit 8.3

The affect of coefficient of variation on sample size. The diagram illustrates the number of times a sample size based on a C of V of 1 must be increased to gain equal accuracy at higher C of V

8.9.2. Critique of the mean-per-unit method

Mean-per-unit, without stratification, is not suited to sampling most accounting populations because of the high degree of value skewness inherent in such populations. This high degree of skewness generates a coefficient of variation which usually lies in the range of 2 to 5 (Neter and Loebbecke 1975).

Such a high coefficient of variation, in turn, will tend to generate large audit sample sizes at any given level of accuracy. Such sample sizes, which may easily run to several thousand units, will be likely to far exceed the sample sizes traditionally drawn by external auditors Exhibit 8.3 illustrates the problem.

Thus the nature of accounting populations makes simple mean-per-unit estimation inefficient,as a sampling technique.

The simple mean per unit method also suffers from several minor defects. For example, to calculate the audit sample size accurately the standard deviation of the audited amount must be known. But this is not known and so it must be estimated from a sample (7). This sample, as we noted above, is likely itself to be highly skew and so the sampler either takes a small sample with a high standard error as his estimate or a large expensive sample. Neter and Loebbecke (1975) found that for typical accounting populations a sample size of at least 200 was needed to estimate the standard deviation to a fair degree of accuracy. The degree of error in estimating the SD is important. Roberts (1978) states that "Any percentage error in estimating the SD results is about twice that percentage error in the calculated sample size". (p.64)

A further problem in applying mean-per-unit estimation to audit work is that, if no errors are found, the mean-per-unit estimate will not in fact be the best available estimate! For example, if the recorded total value is £x and the mean-per-unit estimate £(x + y) and no errors are found in the sample, then £x, not £(x + y) is the best available estimate since the auditor has no grounds for believing £x to be incorrect.

This last point directs attention to that school of thought which believes that an auditor should concentrate on evaluating differences i.e. errors, and not account totals, when recorded values are available.

We conclude that mean-per-unit estimation is only suited to audit work if the population under audit is not properly recorded, contains many errors or has a low coefficient of variation. See Pushkin (1978).

Few accounting populations have these characteristics and so simple mean-per-unit estimation is seldom suited to audit work.

8.10 MEAN-PER-UNIT WITH STRATIFICATION

The earliest writers on evaluating value error (Vance 1948) (Neter 1950) noted the high degree of value skewness inherent in accounting populations and its impact on audit sample size. They, and all subsequent writers on the subject, have emphasised the need to stratify accounting populations under audit to reduce the coefficient of variation within each strata and so reduce audit sample size for any given level of accuracy.

As noted in Chapter 3, the stratification procedure requires the auditor to decide on an optimal number of strata, the location of the strata boundary, the method of allocating the total sample between the strata, and the calculation of sample size given the required degree of accuracy of the estimate.

8.10.1. Stratification technique

When deciding on the number of strata the auditor is faced with a trade-off between cost of stratification and cost of sample size.

The greater the number of strata the smaller the total audit sample is likely to be for any given degree of accuracy in the estimate. However a large number of strata become difficult to handle, even with a computer, since the sample must usually be resorted and screened by hand (8).

Roberts (1978) suggests five strata as an optimal number, but one large audit firm have reported that they use up to twenty strata (9). The auditor must balance the benefit of reducing sample size against <u>his</u> additional cost of applying complex stratification procedures. There is no general solution.

Several methods of deciding on stratum boundaries have been suggested (Arkin 1974) (Roberts 1978). These methods were outlined in Chapter 3. All of these methods remove a top stratum of high value units, every one of which is audited. The estimation procedure applies only to the units remaining in the population after lopping off the top stratum.

The relative efficiency in minimising sample size can vary a great deal between methods. Among the three methods listed in Chapter 3, the 'standard deviation' method can be shown to be the most efficient assuming a high correlation between recorded values and audited values. If, however, this correlation is low, say $>.80$, then the accuracy is spurious, since the estimate of the standard deviation of each stratum will be wide of the mark, and consequently the audited value estimate will also be inaccurate.

The "equal recorded value in each stratum" approach avoids this difficulty. The method chosen will then depend on the auditor's prior estimate of the error rate in the given population. If this is expected to be low, the standard deviation method is likely to be the better choice

Neter and Loebbecke (1975) ran an audit simulation using mean-per-unit with stratification. They sampled four accounting populations with a variety of error conditions.

Their main findings were:

(a) MPU with stratification is more precise than ratio or difference estimation without stratification.

(b) The standard error was less sensitive to error conditions compared to simple MPU

(c) 15 strata appeared about optimal

(d) Nominal confidence coefficients were close to the actual coefficients calculated from the simulation.

8.10.2. Advantages of MPU with stratification

By applying stratification to mean-per-unit sampling, we may be able to effect a massive reduction in audit sample size, assuming the usual value skewness inherent in accounting populations. The technique may also unfortunately introduce the following problems.

8.10.3. Disadvantages of MPU with stratification

If errors are few or highly skewed as to value, the stratified approach is subject to the same criticism outlined above in our discussion of simple mean-per-unit estimation. If no errors are found in the stratum the recorded value is the best estimate (if available) and a precise estimate of the stratum standard deviation may well be expensive in sample size.

The method is also complicated to apply unless the accounting population under audit is stored in a computer and a suitable programme package is available.

The method seems best suited to the audit of accounting populatio n s such as inventory which tend to have high low value error rates

8.11 AUXILIARY ESTIMATORS

Mean-per-unit methods estimate the total value of the accounting population under audit.

This approach, it can be argued, focusses on the wrong objective. The objective of the auditor is to estimate the differences between the recorded values and the audited values. Is it not more direct and also more economical to seek for these differences and extrapolate them to estimate the total difference between the recorded and audited population?

Such a method provides what statisticians call an auxiliary estimator.

In audit work two types of auxiliary estimator have been used. The first is called a difference estimator and the second a ratio estimator.

8.12. DIFFERENCE ESTIMATION

When an auditor uses difference estimation he draws an unrestricted random sample from the population under audit, tests the sample to detect differences between the recorded and audited amounts, and extrapolates the mean of this sample of differences to provide an estimate of the total difference.

The difference can be either an overestimate (recorded amount exceeds audited amount) or an underestimate (audited amount exceeds recorded amount)

8.12.1. Some advantages of difference estimation

Traditional auditing seeks for differences between a sample of recorded values and their audited values. Thus difference estimation follows traditional audit practice. This provides a strong argument in its favour.

Also it can be demonstrated that difference estimation is more efficient than unstratified mean-per-unit estimation if the correlation between the recorded and audited amounts exceeds 50%.

The standard error of the difference estimate equals the standard deviation of the sampling distribution and this standard deviation can be used to estimate the accuracy of the total recorded value. Since this standard deviation is likely to be much smaller in absolute value compared to the SD of the mean-per-unit estimate of identical sample size, the difference estimate can be said to be more efficient than the latter. Finally we should note that the difference estimate becomes more efficient as the variance among the differences discovered decreases. The ideal sample for difference estimation arises when all differences are equal in value.

These characteristics of difference estimation suggests that it may be suited to estimating the degree of value error in accounting populations under audit.

Unfortunately the method suffers from some defects when applied to audit work.

8.12.2. Limitations of difference estimation

The problems in applying difference estimation to audit work arise from the nature of error in accounting populations. In most accounting populations the error rate is too low and the error distribution too skew to permit difference estimation to be applied effectively. The ideal audit situation for applying difference estimation would be as follows (a) a high proportion of non-zero differences would exist in the population say $>$ 20% (b) the non-zero differences would be small and approximately equal in value (c) both under and overstatements would occur, in approximately equal proportion.

Chapter 3 demonstrates that error rates and values in accounting populations have characteristics almost exactly opposite to those stated above. Almost all accounting populations, except inventory, have error rates well below 20%, the differences are frequently large, 100% error being not uncommon, and far from being equal in value, the distribution of error values are invariably highly skew. Finally, every study of accounting errors published so far has shown overstatements to be much more common than understatements. Thus the ideal conditions are unlikely to be realised.

The reasoning behind the "ideal conditions" is as follows:

(a) If the proportion of non-zero differences is less than, say, 20% the probability of picking up a sufficient number of non-zero differences is small. In the UK audit sample sizes fall in the range of 40 to 150. Sampling theory demonstrates that at this size of sample less than 20 non-zero differences will frequently be picked up if the proportion of non-zero errors in the population being sampled is less than 20%. A total of less than 20 differences is insufficient to assume a normal distribution for the difference estimator. See Roberts (1978) p.73.

If the non-zero differences are <u>not</u> symmetric around zero value the situation is even worse, and a lack of symmetry is the norm with accounting errors!

A further reason for requiring at least 20 non-zero differences is that the standard deviation of the total population of differences is not known by the auditor. He must estimate this from the sample. But if the sample contains less than about 20 non-zero differences, the estimate of the SD will be rather imprecise and the SD has a strong impact on sample size. (Roberts (1978) provides a method of estimating SD if the number of non-zero differences is less than 20, using student's 't' table rather than the Normal table).

However these adjustments make the technique complicated to apply. These complications suggest that difference estimation is unlikely to be applied by an auditor to a population where non-zero errors are anticipated to be less than 20%

8.13. RATIO ESTIMATION

If difference estimation proves unsatisfactory as an audit sample method an alternative auxiliary estimator is available in the form of ratio estimation. When an auditor uses ratio estimation procedures, he draws an unrestricted random sample from the population under audit, tests the sample to detect differences between recorded and audited amounts, and then calculates the ratio d/y, where d represents the difference and y the recorded amount containing the error. He then calculates the mean-per-unit of the sample ratios and extrapolates this ratio to the total recorded value to estimate the total difference. (i.e. total value of error).

One minor restriction is that all recorded amounts must be positive, or at least that negative amounts must be segregated and audited independently. The differences can be either overstatements or understatements.

8.13.1. Some advantages of ratio estimation

When discussing difference estimation we noted that the ideal pattern of differences arose when the differences were of equal absolute value. Most empirical studies of error value (see Chapter 3) have concluded that error values tend to be proportionate to the value of the recorded values which generate them. For example many recorded values are 100% tainted (see Chapter 3) and we know that accounting populations are almost all highly skew with regard to value.

Ratio estimation is ideally suited to this condition. Ratio estimation is at its most efficient when the error value is a constant multiple of its recorded value.

For this reason ratio estimatesin audit work will normally be more economical in sample size for a given degree of accuracy compared to difference estimates.

8.13.2. Limitation of ratio estimate

Ratio estimation is more economical in sample size compared to difference estimation for auditing standard accounting populations, the method also enjoys the same advantages as difference estimation over mean-per-unit (10) but, despite this, its usefulness in external audit work is very limited.

The reason for this restricted usage again stems from the nature of accounting and accounting error populations.

The limitations, as might be expected, are similar to those applied to difference estimation.

The proportion of non-zero differences must be high, say over 20%, for a sample of conventional audit size to have a reasonable chance of picking up 20 or more non-zero differences.

Failing this the sampling distribution of the ratio estimator may not be normal and so complications arise which make the method difficult to apply.

Cochrane (1963) states that for the normal approximation to apply, the sample size should be large enough to make the coefficient of variation of both the recorded and audited amount less than 10% when divided by the square root of the sample size. (p.162).

If we apply this rule strictly to audit work where the sample size is traditionally 40-150 the coefficient of variation must be between 0.60 and 1.20. Unfortunately in most accounting populations the coefficient of variation lies between 1.50 and 4.00. Even if the top value stratum is removed and audited 100% the c of v usually falls to between 1.00

and 3.00 which is still high for Cochrane's requirement.

Neter and Loebbecke (1975) have demonstrated that if the non-zero ratios are not distributed symmetrically around zero the sampling distribution is unlikely to be normal even when a large number of differences are detected, say in excess of 50. Empirical studies (see Chapter 3) show that non-zero differences are almost always distributed assymetrically.

Two further problems should be noted. First, as with difference estimation, it may be difficult to achieve a very precise estimate of the standard deviation of the population of non-zero ratios. Second the mean value of the total population of all possible samples of ratios does not equal the audited value exactly, although the difference is small if the sample is of traditional audit size, say > 50.

These limitations suggest that although ratio estimation is better suited to external auditing than difference estimation, its use is restricted to auditing accounting populations with high error rates in excess of 20%

We conclude that neither difference estimation nor ratio estimation is suitable as a sampling method to be applied generally in external auditing.

8.14. AUXILIARY ESTIMATORS WITH STRATIFICATION

Mean-per-unit estimation generated sample sizes too large for audit work until stratification was applied to the accounting population. Stratification reduced the variance within each stratum and so reduced the required sample size at any given level of accuracy. Can a similar approach be adopted to make difference and ratio estimation more suited to audit work?

The major problem with mean-per-unit estimation was the high coefficient of variation. This can also be a problem when making difference and ratio estimates but it is not the main problem.

The main problem with difference and ratio estimates is the small number of non-zero differences likely to be found in an audit sample of conventional size (40 - 150) if the error rate is below 5%, which it usually is. Stratification does nothing to solve this problem therefore its contribution to making auxiliary estimates more amenable to audit work is rather limited.

8.14.1. Stratified difference estimation

This method is similar to unstratified difference estimation except that difference estimates are made for each stratum and these are added together to arrive at an estimate of the total difference. This should provide a smaller sample size at any given level of accuracy.

Since very few and often no differences are found in some strata it is difficult or impossible to calculate the standard deviation of the stratum differences with a reasonable degree of precision. If the SD of recorded accounts is used as a proxy this will give a conservative estimate so long as the correlation between recorded and audited amounts exceeds 50%.

The usual basis for stratification will be the recorded amounts. If this is so, the sampling distribution will not be even approximately normal unless the error rate exceeds 20% or the non-zero differences are relatively small compared to the average value of the stratum. See Neter and Loebbecke (1975). In Chapter 2 we showed that these conditions are unusual in accounting populations.

The other limitations which we noted with regard to difference estimation also apply to stratified difference estimation.

We conclude that stratified difference estimation is not suited to general audit work although the method may be appropriate to specific audit problems where many ($\succ$ 20%) small differences occur and an overall estimate of total difference is required.

8.14.2. Stratified ratio estimation

If stratification imposes little improvement on difference estimation can it improve the auditing efficiency of ratio estimation? Two methods of applying stratification to ratio estimates are available, the first of these is called combined ratio estimation and the second separate ratio estimation.

Combined ratio estimation

This estimate of total audited amount is formed by using the formula

$$X = \frac{A}{R} \times T$$

where X = combined ratio estimates of total audited amount

A = stratified mean estimate of audited amount

R = " " " recorded "

T = total recorded amount

The combined ratio can be used to make an estimate of the total value of error.

This method enjoys one considerable advantage over stratified difference estimation. This advantage is that the combined ratio estimate does not require several differences to be present in each stratum. The estimate is not unduly sensitive to zero errors in a stratum.

However, the infrequency of non-zero value errors in accounting populations is still a problem, although rather less so than with stratified difference estimation.

The standard deviation of the stratum differences must again be estimated.

As with difference estimation, the SD of stratum recorded amounts is usually taken as a conservative proxy for audited amounts. An alternative possibility would be to use the stratum SD calculated in a previous year.

The actual shape of the sampling distribution of the estimates is again a problem. Neter and Loebbecke (1975) have experimented with simulations over a range of possibilities. They conclude that (a) the bias of the ratio estimator with samples over 100 units is negligably small. (b) the standard error of the stratified ratio estimator is almost the same as the stratified difference estimator. (c) the difference and ratio estimators are more precise than the mean-per-unit estimator when stratification is employed to all three methods. (Page 94).

However, the simulation throws grave doubt on the reliability of the nomin al confidence intervals claimed when using stratified ratio(or difference) estimation, if the proportion of non-zero differences is less than 5%. (Page 101-102).

Neter and Loebbecke show that "for low error rates (5% and below) the proportion of correct confidence intervals for the stratified difference and ratio estimators (based on sample size of 100) are so far below the nominal confidence levels as to render the latter meaningless!! Even at an audit sample size of 200 the nominal confidence intervals are unsatisfactory.

We conclude that when using these methods conventional calculation of confidence levels using the normal approximation techniques cannot be used for auditing accounting population s with error rates anticipated to be below 5%.

Since most accounting population will fall into this category these methods are not suitable for general use in auditing. They may be helpful in specific cases with high error rates (but see Neter and Loebbecke page 102 where even a 30% error rate is too low to use the method!)

Separate ratio estimation

Using this method the ratio estimator is calculated by adding together "the sum of each stratum's ratio times its population recorded amount." (Roberts 1978 P.109). Roberts claims that the method will provide a more precise estimate if the ratios for the various strata vary widely. Unfortunately this method suffers from the same major defect that limited the application of stratified difference estimation. Several non-zero differences are required in each stratum. Since this condition will not apply if the error rate is low ($<$ 5%) the method is inferior to combined ratio estimates in most situations.

Separate ratio estimation is only suited to audit sampling if the anticipated error rate is high ($>$ 20%) and a wide disparity is anticipated between the various stratum ratios.

These conditions may arise, say in inventory audit, but they are unusual.

8.15. CONCLUSIONS RE AUXILIARY ESTIMATORS

We conclude that auxiliary estimators with or without stratification, are not generally applicable to external audit work.

The major obstacle to their use is the small number of non-zero value errors which occur in most accounting populations.

This low rate of value error makes it difficult to calculate the standard deviation of differences and provide a sample large enough for the sampling distribution of the estimate to approximate to a normal distribution.

These problems can be overcome but not without undue complication which makes the methods unsuitable for general audit use.

8.16. MONETARY UNIT SAMPLING

The system we have called monetary unit sampling (MUS) is the sampling method currently most widely used to evaluate error in accounting populations. Our survey (Chapter 6) suggests that over 90% of audit applications of SS which attempt to evaluate error value use some form of MUS. As noted in Chapter 6 several versions of the method are extant. The most important distinction between each version being the method used to calculate the upper bound on the estimated value of error in the population under audit.

The basic approach, as was explained in Chapter 4, is as follows:

(1) Differentiate between audit unit and monetary unit

(2) Divide the audit unit in the population under audit into the three categories of positive, negative and zero balance.

(3) Decide on an acceptable upper error limit (This is normally derived from the materiality value) and degree of reliability required

(4) Use the Poisson table to calculate the required sample size and so the skip interval

(5) Draw the monetary unit sample and identify the number of errors and degree of tainting of each error.

(6) Calculate an upper bound on the estimated error value and decide whether this is acceptable.

Our study to date has found only the mean-per-unit sampling method with stratification suitable for the purposes of external audit work. The popularity of the MUS technique suggests that this method must also be a strong contender for general application.

In the following section we will explain why many experts on SS claim that the MUS method is the sampling method best suited to estimating error value in accounting populations.

In a later section we will carefully evaluate the various criticisms which have been raised against the MUS method.

8.17 THE ADVANTAGES OF MUS

The MUS method enjoys several powerful advantages over other methods of scientific sampling applied to audit work. The advantages can be assigned to two classes, statistical advantages and operational advantages.

8.17.1. Statistical advantages of MUS

1. **The method is distribution free**

   All the variables sampling methods described above suffer from the limitation that the sampling distribution is assumed to approximate to a Normal distribution. If this assumption is invalid the nominal confidence levels attributed to the estimate may be incorrect Neter and Loebbecke (1975) provide strong evidence in support of this criticism from their simulation of variables sampling from highly skewed accounting populations.

   The MUS method makes no assumptions about the shape of the sampling distribution and so is freed from the Normality assumption. MUS is distribution free.

2. **Distribution of error sizes can be ignored**

   The population error values estimated from a sample can be affected by the distribution of total error value within the recorded population. The total error value can be contained in a single large error or a large number of small errors. This distribution of the total value of error affects variable sampling estimates rather severely since a few large errors might be missed if they are not individually material.

MUS sampling can be so designed that a "worst" distribution of error value is assumed. This means that no reliance need be placed on a probable distribution of error sizes or of item values.

3. <u>The upper bound a conservative estimate</u>

The MUS method calculates an upper bound on the estimate of value error remaining in the accounting population under audit.

Several studies have proved or at least demonstrated that this upper bound is conservative for low error rates ($<$ 5%) so far as beta risk is concerned. Feinberg, Neter and Leitch (1977) provide a mathematical proof and Gartstka (1977) and Neter and Loebbecke (1975) ran simulation models which reached the same conclusion.

The exceeding conservatism of some bounds, for example the Stringer bound, may in fact be so severe as to be a positive disadvantage.

We will return to this point later.

Note that bounds can be devised for MUS which are not conservative and the conservative bias does not apply to high error rates ($>$ 10%) See Neter and Loebbeck (1975) Chapter 9.

4. <u>The evaluation of several populations can be combined</u>

The conventional audit consists of several accounting populations.

If a separate bound is calculated for each population it is possible to combine these populations to calculate an overall bound for the total population.

If variables sampling is used it is difficult to combine the several estimates of error into a single overall estimate.

5. Sub-sampling of combined audit unit is easy

It frequently happens that a primary audit unit is made up of several sub-units. For example, a total contract bill (the primary audit unit) is made up of several bills from sub-contractors. The MUS method is ideally suited to handling this type of sampling unit since the individual sample will be drawn from a sub-contractors bill and this single bill can be audited. A variables approach requires a two step evaluation process.

6. MUS is technically simple

Although the theory behind the MUS method is not particularly easy to understand the calculation of the 'skip' interval and the evaluation of the sample results are easy to effect once suitable tables are provided.

This is not true with regard to any of the variables sampling techniques. Both the theory and the practical implementation are complicated enough to require the use of a computer.

8.17.2. Audit advantages of MUS

MUS was specifically designed to handle audit sampling of accounting populations. The audit advantages built into the technique are as follows:

1. MUS follows traditional audit procedures

For example, MUS increases the probability of testing high rather than low value audit units and it requires the population under audit to be added up.

2. The materiality figure for the given audit can be directly plugged into the MUS sampling evaluation procedure

The upper bound on the estimated error value can be directly related to materiality.

Various bounds have been suggested (see ahead) but whichever is used materiality can be related to it.

Either the 'estimation' approach or the 'hypothesis testing' approach can be used depending on whether the audit firm wishes the audit to be judgemental or decision orientated.

Kaplan (1975) provides a method of relating materiality to the upper bound on estimated error value.

The usual procedure is for the auditor to specify materiality £M in £ sterling terms and then subtract from this figure the most likely error value £E, anticipated for the audit as a whole.

£(M-E) represents the tolerable upper bound on error value. See Teitlebaum etc., (1975).

3. <u>A computer is not essential for operating this method of estimating an upper bound on error value</u>

Variable methods using stratification require a computer to identify strata and evaluate the sample results. MUS sampling procedures can be operated without the need for a computer.

A leading practitioner of MUS, Clarkson Gordon, Canada, use manual methods on about 60% of their MUS audits. See Teitlebaum (1975) p.39.

AAFI, a leading US association of accountants, employ the manual method almost exclusively.

Although a computer is not a necessary condition for using MUS, if a computer is available the advantages provided over the manual method are considerable (McRae 1980). Several computer programme audit packages such as DUSPAK and AUDITAPE incorporate MUS.

While on the topic of computers, we should note that MUS can be performed as a <u>single</u> run during the client's tape processing of a regular job. Most variable sampling systems require at least <u>two</u> runs of the tape, which can prove an expensive operation.

4. <u>MUS is designed to pick up the normal level of error rates found in accounting populations.</u>

As we noted above, MUS reliability levels tend to be conservative when error rates are low ($<$ 5%). This is the usual condition on accounting populations as noted in Chapter 3.

Other estimating procedures, for example auxiliary estimators with stratification, work well for high error rates ($>$ 20%) but not for the levels usually found in accounting populations excluding inventory.

MUS also handles highly skewed error values better than alternative methods since a bound based on the <u>worst possible distribution</u> of error value can be calculated.

5. <u>The sample sizes generated by MUS when using reasonable criteria for materiality and reliability are within the range of traditional audit sample sizes.</u>

There can be no doubt that this statement is true when applied to external audits in North America. The normal sample sizes in North America are around 100-350 units per population.

In the UK the traditional sample sizes are lower, being of the order of 40-150, with a model value around 60.

It is possible that in order to meet these lower audit sample sizes the conservatism inherent in current MUS methods may have to be squeezed out: Gartstka (1977) Mottershead (1980) and McCray (1980) have all suggested methods of calculating less conservative upper bounds.

8.18. The case against MUS

The previous section described the many arguments in favour of adopting MUS as that sampling system best suited to estimating an upper limit on error value existing in accounting populations under audit.

The MUS system has, however, not been immune to criticism in the accounting literature. We will now examine this body of criticism to see if it seriously undermines the credentials of the MUS method. We will pay particular attention to points which affect the practical application of MUS.

We emphasise relevance to practical application since many criticisms seem of limited significance in the context of practical auditing. The arguments against MUS are of varying weight. We will set them out in what we believe to be reverse order of importance.

The criticisms have been culled from an extensive search of the literature on MUS. The main critics of MUS have been Goodfellow, Loebbecke and Neter (1974), Gartstka (1977), Neter and Loebbecke (1975) and Roberts (1978).

1. The Poisson distribution is unsuitable for calculating sample size.

   Most MUS systems use sampling tables based on the Poisson distribution for calculating sample size. See for example Leslie etc. (1980) and the DHS Audit Sampling Manual (p.57).

   From a statistical viewpoint the hypergeometric distribution is the most suitable distribution since audit sampling consists of drawing sample units without replacement from a finite

	No. of errors	Poisson	Binomial	Hyper G
		Sample Size		
(1) N = 8309	0	167	166	166
£Y = £379,131	1	265	263	263
M = 2%	3	432	430	430
(2) N = 1000	0	20	19	19
£Y = £100,000	3	52	50	50
M = 15%	6	79	76	76
(3) N = 200	0	60	59	58
£Y = £1000	1	95	93	92
M = 5%	3	155	153	150

N = number of audit units
£Y = recorded total value of account
M = materiality as % of £Y

Exhibit (8.4) Audit sample size required using MUS Method (DUS type) under varying conditions. Note that a high materiality (2) or small population (3) has little effect on the difference in sample size.

(Adapted from Teitlebaum etc., 1975) p.37)

population. It is, however, a very lengthy arithmetical process to calculate sample sizes from the hypergeometric distribution. (22).

Failing the hypergeometric, one might be tempted to try the binomial distribution, since 'n', the sample size, is usually an insignificant fraction of 'N' the population size, in audit sampling. When 'n' is an insignificant fraction of 'N' the sampling process without replacement is almost identical to sampling with replacement. Therefore, the binomial can be used as a proxy for the hypergeometric. But tables of the binomial suited to audit work are not easy to find and would be bulky, since the binomial is determined by two parameters 'n', the sample size and 'E' the error proportion.

The Poisson distribution is rather easier to handle for computation work.

Poisson probabilities can be exhibited on a single page if the number of errors likely to be discovered are not too high, say less than 50. This is the usual situation in audit work. Thus most practitioners of MUS have used the Poisson distribution to calculate tables of audit sample size.

The viability of using the Poisson as a proxy for the binomial has been examined by Anderson and Burstein (1967). They conclude that the Poisson approximation of the binomial is always conservative and efficient when E is small. This means that at any given confidence level, at a fixed precision, the Poisson will require a larger sample size than the binomial

In an audit context this means that the Poisson generates a rather larger sample than actually required, given the reliability and precision stipulated by the auditor.

Is this excess audit sample large enough to be of economic consequence to the auditor? GLN (1974) p.16 suggest that it may well be.

Exhibit (8.4) illustrates the sample size required under various postulated error rates at a reliability of 95%. The sample sizes are calculated using the Poisson, the binomial and the hypergeometric distributions.

We conclude that within the range of error rate and sample size usually encountered in external auditing the sample size difference is <u>not</u> of economic significance.

This particular criticism of the MUS technique can be discounted.

2. <u>The MUS system will underestimate the error rate if most errors occur in small value audit units.</u>

Since the probability of selection of an audit unit is roughly proportional to its value, small value audit units will have less chance of selection than large ones. If internal control concentrates audit effort on checking high value units it might well be that more errors slip through the system in low value rather than high value units.

Studies of accounting error have not discovered a higher probability of error in low value units. See for example, Ramage, Kreiger, Spero (1979).

But even if the statement was true would it make a significant difference to error evaluation using MUS?

It would make no difference to evaluating an upper bound on error value using the DUS method since, as we noted above, the distribution of total error value does not affect the conclusions drawn from DUS evaluation (12).

It would, however, affect a calculation of an upper bound on error rate, if MUS were used to assess error rate among audit units rather than error value among monetary units.

We conclude that it is safer not to use MUS to evaluate error rate. If an auditor wishes to evaluate error rate it would be wiser to use acceptance, sequential or survey sampling as outlined in the first part of this chapter.

The above comment refers to overstatements of recorded amounts. Small value units could conceal large understatements. We will discuss this problem under point (7) ahead.

3. MUS sample values cannot be used to estimate the total value of the audited population.

This is a variation on the previous criticism.Since all audit units do not have an equal chance of selection the sample mean value cannot be projected as an estimate of the total audit value. (But see note 12(a)).

The relevance of this criticism lies in audit philosophy rather than statistical theory. The MUS system tests to see whether the error value in the recorded population is significant. If it is significant the recorded value is adjusted, or perhaps rechecked by the client and retested by the auditor. (See DHS "Audit Sampling" p.7)

The audit objective is to ensure that the post audit recorded value is not significantly different from the true value (13). An estimate of the audited value as obtained by variables sampling methods is not required. If the auditor believes that an estimate of audited value is essential, he must use a variables sampling procedure.

This latter approach conflicts with traditional audit sampling method which follows the MUS approach. An estimate of audited value is _not_ projected in traditional auditing. The recorded value is only adjusted if the 'traditional' non-statistical sample suggests that the recorded value is incorrect by a significant amount.

We conclude that the MUS approach to evaluating the total recorded amount is acceptable since it follows the traditional approach in adjusting recorded amounts for error _only_ if the difference is considered _material_.

4. _The MUS method can indicate that a significant value error exists but the method cannot indicate the magnitude of this error. Therefore the auditor is given no help in estimating the magnitude of the required adjustment for error._

Variables estimation methods can provide this information.

The MUS method provides an upper bound on the likely error value. It does not provide a true point estimate of the most likely error value or a lower bound on this estimate.

The criticism, therefore, is valid.

In those few cases where the upper bound exceeds materiality the auditor must either devise some heuristic for estimating the error adjustment or switch to variables estimation.

Teitlebaum etc. (1975) p.14, addresses himself to this problem. He suggests a most likely error estimate based on the errors actually discovered in the MUS sample. If e% of the monetary value of the sample is in error then £(e x Y) is the most likely error value in the recorded population. The estimated value of error £(e x Y) can now be used to adjust the recorded value.

This approach is rather unsatisfactory and more research is needed on the problem.

Regarding variables estimation of error value. It is true that a point estimate and upper and lower confidence bound are calculated, but at the audit sample sizes used in the UK these are likely to be very far apart at 90% reliability or above. The comparative advantage of this confidence interval over the MUS upper bound may thus be somewhat spurious.

The kernal of this criticism is that the size of audit samples traditionally drawn in auditing are large enough to detect the existence of a significant error but not large enough to make a very precise measure of its magnitude.

5. In order to use MUS the population under audit must be added up

The MUS method was designed to fit in with traditional auditing procedures. One traditional auditing procedure is to add up the population under audit. Therefore the incremental cost of adding up the population for the purpose of selecting the MUS samples will often be close to zero. Also, if the population under audit is being processed on a computer, the "adding up" can be performed rapidly and at low incremental cost (a few £s) by the computer. If neither of these conditions apply several short cut methods of adding up have been devised by practitioners of MUS. See Leslie, Teitlebaum, Anderson (1980).

Failing this an alternative sampling system which does not require the population to be added up has been devised. This method is based on a two stage approach. The first stage samples physical units the second stage monetary units. The method can reduce the cost of adding up required by 50% to 90%. See Roberts (1978) p. 22 and Raj (1968) (14). We doubt whether the adding up problem presents a serious limitation on the application of MUS.

6. A full mathematical proof of the MUS method has not been worked out.

A full mathematical proof of the Stringer-Stephan system was never provided by the authors. An article by Feinberg, Neter and Leitch (1977) sets out a mathematical exposition of the Stringer-Stephan system together with other variations of the MUS method. Goodfellow, Loebbecke and Neter (1973) set out a mathematical exposition of their CAV system in appendix B and C. Teitlebaum (1973) and Aldersley and Teitlebaum (1979) provide a mathematical exposition of the DUS system. Roberts (1978) Chapter 6 provides a mathematical model of the PPS system for both low and high error rates and Kaplan (1975) and Garstka (1977) provide further mathematical analysis of their proposed system for estimating upper bounds on error value.

It is true that these expositions do not provide a rigorous proof of the MUS method but simply prove that the various bounds are conservative for beta risk, the more important sampling risk in audit work.

The lack of a rigorous mathematical proof is likely to be of more concern to the statistician than to the auditor. The auditor is likely to be more concerned with the question "does it work?", than the question "why does it work?".

Neter and Loebbecke (1975) Garstka (1977) and Reneau (1978) have run extensive simulations to test various MUS bounds under a variety of different audit conditions. They conclude that the method works, i.e. the bound is conservative, so long as the rate of error is low, say, less than 5%.

We are informed that D. Roberts of the University of Illinois is currently supervising a further series of tests. John McCray has run a simulation to test the Stringer bound (21).

Since the more commonly used versions of the MUS system appear to work as predicted under simulated conditions we doubt whether the absence of a full mathematical proof is likely to deter an auditor from using it.

7. <u>MUS cannot handle zero value balances or understatement errors</u>

Since audit units are selected under MUS by hooking an individual monetary unit, i.e. a £, an audit unit of zero value has no chance of selection. This weakness does not apply to variables sampling where an account of zero value has an equal chance of selection with any other account.

Auditors using MUS must perform an additional sampling operation to test zero balances. If the number of zero balances can be calculated using a computer, or estimated using a manual method, attribute sampling theory can be used to calculate a sample size sufficient to test that at least x% of the zero balances are in fact zero at a given level of reliability. If the zero balances are not zero they will almost certainly be understatements.

The problem of understatements has always posed a major problem for auditors. An understatement occurs when the audited amount exceeds the recorded amount. Unless the balance is negative, which is unusual, the value of an overstatement is limited to the value of the recorded amount. The value of an understatement has no such maximum limit.

In variables sampling understatements can be deducted from overstatements and the net balance used to extrapolate the total value. This simple approach cannot be employed with MUS which is an attribute sampling method. A separate calculation must be made to estimate a bound on overstatement and understatement. These estimates may then be netted off against one another in some way. Goodfellow, Loebbecke and Neter (1974) are particularly critical of the method by which MUS handles understatements. "The most basic deficiency from an audit point of view appears to be the imbalanced treatment of errors of over and understatements" (p.23). GLN are referring to the Stringer -Stephan method of treating understatements. Stringer-Stephan calculate an upper bound on overstatement error and a lower bound on understatement error and subtract one from the other to arrive at an upper bound on the net overstatement error. This approach assumes that not understating overstatement error is the audit objective. In this sense it is very conservative and this conservatism could result in substantial overauditing.

On the other hand we should note that if the reliability attached to both bounds is R%, the reliability attached to the net bound will be less than R%. It will be between (2R-1) and R. See Roberts (1978) p.125.

The DUS system uses a different approach in handling understatements. The method is as follows:

(a) An upper bound on both understatement and overstatement error is calculated.

(b) The upper bound in both cases is then reduced by the most likely error in the opposite direction.

(c) The 'net' bounds are both studied for likelihood of breach of materiality (see Teitlebaum 1973 for a mathematical analysis of this approach)

This latter approach presents a more balanced treatment of over and understatements. It is a good deal less conservative than the Stringer-Stephan approach.

Empirical studies of accounting errors, such as displayed in Chapter 3, indicate that large understatement errors are rare. Certainly users of accounts tend to worry more about overstatement of recorded amount rather than understatement.

Since both the Stringer bound and the DUS bound are conservative (for low error rates) in the sense of tending to overestimate overstatement error this criticism is only a problem if

(a) large understatements are likely or

(b) the excess conservatism results in overauditing.

The latter problem will be discussed later. We have already noted that large understatements are rare.

Finally we should note the defence against understatement error which has often been put forward by auditors, namely that a reciprocal population will often exist where the error is an overstatement.

This reciprocal population can be audited by normal means.

Other statistical methods have been devised for detecting understatements, such as using regression estimates or the minimax method used by DHS.

In our opinion the DUS version of MUS handles understatements sensibly, but if the risk of understatement is thought to be a major problem than an alternative statistical approach such as regression analysis should be employed.

Fraud, of course, involves large understatements but we have specifically excluded fraud from this analysis since it is the subject of a parallel study.

8. <u>The nominal confidence levels provided by the MUS system on the estimate of the upper bound of error value are not accurate.</u> The nominal confidence level may be too high (overauditing) or too low (underestimating beta risk).

Neter and Loebbecke (1975) ran a simulation of 600 upper bounds from a variety of accounting populations using the CAV version of MUS. They found that "the actual proportion of correct intervals <u>exceeded</u> the nominal confidence coefficient of 95%" on all populations examined. This means that the MUS bound on this particular version of MUS was conservative in all cases.

The Garstka (1977) simulation also concluded that the DUS version of MUS was almost always conservative.

The Feinberg Neter Leitch (1977) analytical study of MUS bounds concluded that the MUS bounds studied were almost certainly conservative in all cases.

If the bound is conservative the risk involved is the lesser alpha risk of overauditing, not the more serious beta risk of accepting a population containing material error.

We were only able to find one reference in the literature which accused the MUS system of <u>underestimating</u> beta risk. This was implied in Goodfellow, Loebbecke and Neter (1974) when they accused the DUS version of MUS of understating the upper bound on overstatement error. (p.24) In an incisive and detailed reply Teitlebaum etc., (1975) demonstrates that GLN are wrong in claiming that the <u>distribution</u> of total error between audit units can affect the accuracy of the DUS estimate of the upper bound on error. The gist of his argument is that the DUS version of MUS assumes the worst possible distribution of error value so far as estimating the upper bound is concerned. The largest fraction of the Poisson increment is charged with the highest error tainting (100%), the next highest Poisson increment with the next highest tainting,...... and so on. Thus the upper bound calculated is the worst possible given the errors discovered. Again the bound is conservative for beta risk. We conclude that the upper bounds calculated on error value using the MUS system are only likely to be inaccurate to the extent that they are conservative, i.e. that they overestimate overstatement error (15). This type of conservatism would normally be acceptable to an auditor so long as it is not bought at too high a cost.

9. The sample sizes generated by the MUS system of audit sampling are too large to be acceptable to auditors in the United Kingdom.

Chapter 6 demonstrates that three large acounting firms in the UK are using MUS extensively in their audit work, three other large firms and a good number of medium sized firms are using MUS on large audits.

In the USA many large and small firms are using MUS.

Such evidence could, by itself, be considered sufficient to refute the above criticism.

The comment, however, requires deeper analysis since it is the one most frequently levelled against MUS.

There can be little doubt that audit sample sizes in the UK are lower, on the average, than in North America. We estimate the modal sample in the UK at around 60 and in the USA at around 120.

The sample size required to constitute an MUS sample is determined by two parameters the materiality value and the level of reliability demanded

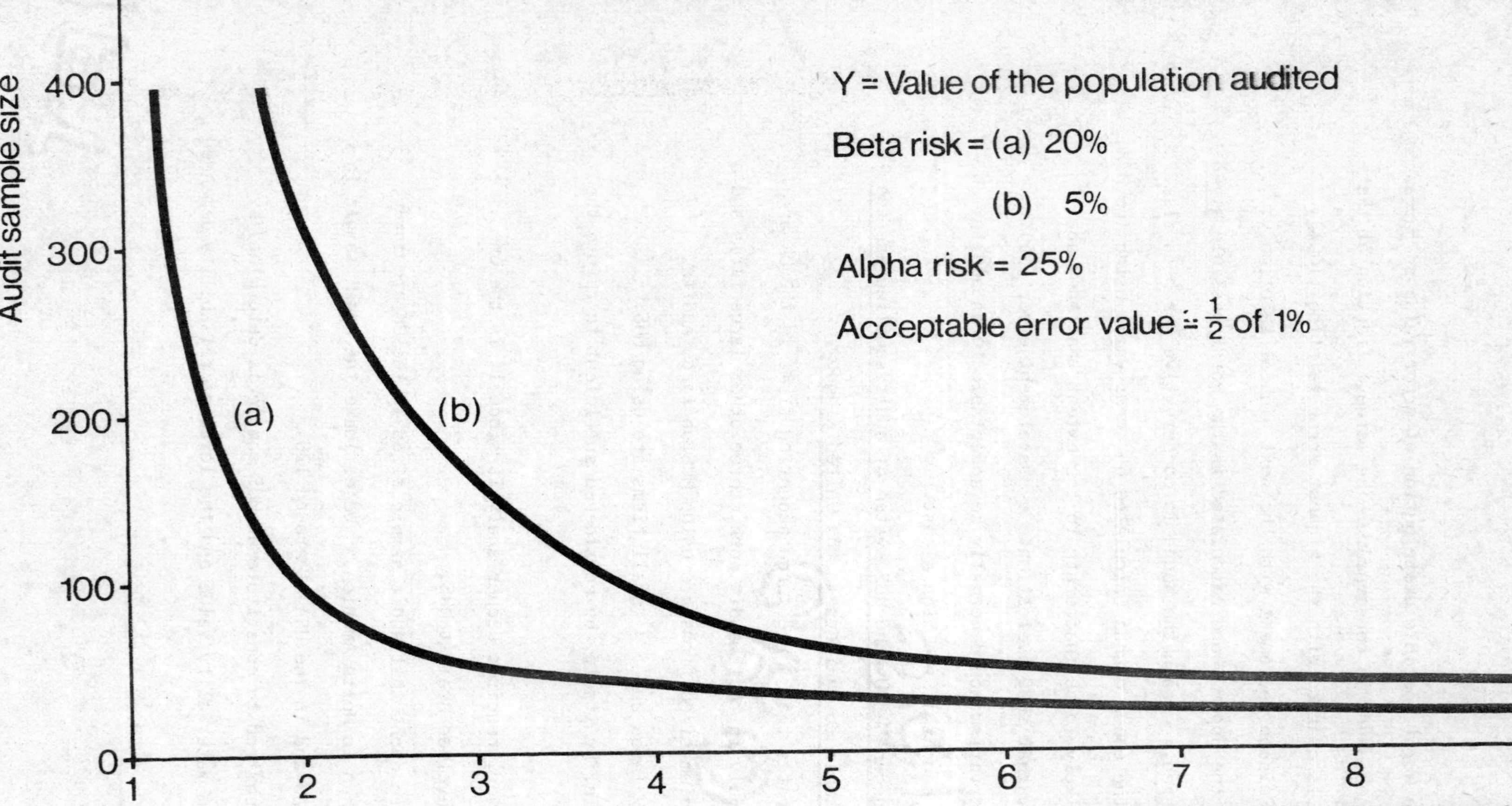

Exhibit 8.5. Audit sample size at various levels of materiality. The sample size has been calculated using the MUS system. Sample size calculated from Kaplan's Table. See Roberts (1979) Appendix 6A, and Appendix to this Study.

by the auditor in the inference from the sample. It is, therefore, of some relevance to calculate the minimum upper bound on the estimated error value which can be provided at various reliability levels for samples of between 50 and 300.

We can use the MUS "acceptance sampling" tables provided by Roberts (1978) appendix 6a, based on Kaplan (1975) to calculate the sample sizes generated by various levels of materiality. The results are shown in Exhibits (8.5) and (8.6) We assume that the following are held constant

(a) Alpha risk of 25%

(b) Acceptable error value at ½% (16) of the value of the population being audited

(c) Beta risk of 5% and 20%

(d) Materiality varying over a range from 1% to 9% of the value of the population being audited.

The Kaplan tables are useful for the purpose of the present discussion since they allow us to take both alpha and beta risk into consideration. Exhibit (8.5) demonstrates that at a beta risk of 5%, materiality would be around 5% of population value for a sample of 50 and 3% for a sample of 150. At a sample size of 300 materiality could be reduced to about 2% of the population value.

These levels of materiality might be considered high for the UK.

Exhibit (8.5) illustrates what happens to materiality under MUS if we use a Bayesian approach and increase our beta risk to 20% (80% level of confidence) at the same size of sample.

The results are as follows:

Sample Size	Approximate Materiality % of Y
20	8.0
60	2.5
150	1.5
300	1.2

We doubt whether alpha risk can be increased much above 25% or beta risk above 20% (17). The acceptable % of ½% could, perhaps, be reduced but overall it is difficult to see how materiality can be reduced much below 1.5% of the population value if audit sample size is to be kept below 150. We conclude that in the UK, but not in North America, the minimum level of materiality using classical sampling techniques is around 3% of the value of the population being audited. Bayesian techniques can reduce this % to around 1.5%. This level of materiality should surely suffice? (18) Current research on MUS is aimed at reducing audit sample size by squeezing out the conservative element in estimated precision.

10. <u>MUS ignores alpha risk. That is the risk of rejecting acceptable populations</u>.

It is true that surprisingly little is said about alpha risk in much of the literature of MUS. Teitlebaum etc., (1975) implicitly consider the point when replying to the GLN (1974) criticism that DUS sampling plans are designed on the assumption that no errors exist, i.e. that DUS is a form of discovery sampling.

Teitlebaum etc., argue that practitioners of DUS can calculate tolerable basic precision in the following way:

	£
Materiality	x
Expected error value	e (from past years)
Precision gap widening	g
Tolerable basic precision	£x - (e + g))

This allows the population to be accepted even if some errors are found. It appears, however, that g is an entirely subjective estimate.

In private conversation D.Leslie, an experienced user of DUS (19), has stated that alpha risk has proved no problem in practice since few populations are rejected by DUS. We infer from this that the proportion of error value is very low in these accounting populations.

Kaplan (1975) has attempted to solve the alpha risk problem (when using MUS) by devising a set of tables similar to acceptance sampling tables. These place MUS in a decision type context as with acceptance sampling. The auditor specifies alpha risk, beta risk, materiality and the expected error value in the population and the tables provide the required sample size and maximum acceptable number of monetary errors. However since all undiscovered monetary errors are assumed to be 100% "tainted" the sample size is likely to be larger than it need be.

We conclude that where the proportion of value error is low, say below 2%, the alpha risk of overauditing is not likely to be high. If the proportion of error is higher, say $>$ 5%, the rejection of too many acceptable populations is a real possibility.

The use of Kaplan's table (or more likely some modified version of it) may point the way towards a solution of this problem.

See Vanecek (1978) for a detailed discussion of this problem.

. The MUS method is too conservative.

Every study of the MUS approach, both analytical and simulative has concluded that, under normal error conditions, all versions of the MUS approach are conservative. See Garstka (1973), Neter and Loebbecke (1975), Neter, Leitch, Fienberg (1978) Vanecek (1979) etc. Some versions are more conservative than others.

Conservatism as to degree of risk is normally considered a virtue in auditing (See Sneed 1979), so there is no harm in this characteristic of MUS so long as it is not bought at too high a price in overauditing.

Vanecek (1979) studies the problem in some detail and suggests that some sort of offset procedure, such as used in CMA or DUS is essential to reduce the conservatism of the beta risk estimate (p.266). Otherwise a Bayesian approach must be adopted. (p.258.)

The key consideration in answering this criticism must be that practitioners of MUS do not appear to reject too many "acceptable" populations.

8.19 Conclusions on the MUS Method

MUS fits well into traditional auditing methodology, is easy to use and provides answers in a format suited to audit work. The sample sizes generated are within the limits traditionally drawn in North America but not, perhaps in the UK. The sample size problem can be solved if a Bayesian approach is used to determine reliability, and materiality is not too tightly set.

We have carefully examined each of the arguments put forward against MUS and have found none which, to our mind, delivers a mortal wound. There can be no doubt that the efficiency of the MUS method will be improved in the future.

8.20 The efficiency of audit samples used for estimating monetary error value.

We can now apply the criteria of audit sample efficiency devised in

Criteria	Type of Sampling						
	Traditional	M.p.U. with Strat	Difference Est. No Strat	Difference Est. With Strat	Ratio Est. No Strat	Ratio Est. With Strat	MUS
epresentative	N	Y	Y	Y	Y	Y	Y *
orrective	Y	N+	N	N	N	N	N
reventive	N	Y	Y	Y	Y	Y	Y
rotective	Y	Y	N	Y	N	Y	Y
onsistent	N	Y	Y	Y	Y	Y	Y
perationally imple	?	N	Y	N	Y	N	Y
conomic							
a) Weak	Y	Y	N	N	N	N	Y
b) Strong	N	Y	Y	Y	Y	Y	Y

Exhibit (8.6). Criteria for deciding on the relative efficiency of various audit sampling methods.

- value sampling

* representative of monetary values.

Chapter 2 to evaluate the various sampling methods suggested for substantive testing of audit value. See Exhibit 8.6.

We exclude mean-per-unit without stratification for the reasons outlined above.

Mean-per-unit with stratification enjoys the normal benefits accruing to a sampling method based on random sampling. It provides a sample which is representative, preventive and consistent. It is not corrective since it does not specifically seek out errors. However, the concentration on large value items should improve the proportion of total value of error corrected (20).

The mean-per-unit system satisfies both the strong and weak economic criteria. It satisfies the required conditions of accuracy at minimum sample size and the size of sample generated by normal audit conditions falls within the sample size limits traditionally drawn by auditors in the U.K.

The one drawback of mean-per-unit sampling with stratification lies with the difficulty of computation. A computer is needed to calculate the stratum boundaries, draw the sample and calculate the bounds on the estimate.

However, if a computer <u>is</u> available mean-per-unit with stratification must be considered a serious contender for the title of best sampling procedure for substantive testing of value.

Difference and ratio estimation without stratification are so similar that we will consider them together.

They enjoy the usual benefits of producing samples which are representative, preventive and consistent. The samples are not corrective and are not biased towards large values. Operationally the methods are not too complicated and can be performed by hand. The strong economic criteria is satisfied since the methods guarantee the lowest sample size to meet the required conditions on the average.

However, to achieve reasonable accuracy the sample sizes are likely to be much too large to be economic in the context of auditing. The methods, therefore fail the weak economic criteria.

The same criteria apply to difference and ratio estimation with stratification as without except for the protective and operational criteria. Difference and ratio estimates with stratification are protective, since the sample is biased towards units of high value. They are, however, undoubtedly complicated in application and a computer is needed to assist with the calculations.

The weak economic criteria may be satisfied but this will depend very much on the distribution of error value within the population. The nominal confidence limits may well be inaccurate. Note, for example, the findings of Neter and Loebbecke (1975) Chapter 8.

The monetary unit sampling methods are based on a monetary rather than a physical audit unit. Bearing this qualification in mind the MUS sample is representative, preventive and consistent. It is perhaps the most protective of all the methods reviewed. It has also been argued that it provides the best basis for estimating total error value for the purpose of correcting the accounts (Teitlebaum etc. 1975)).

The MUS method is operationally simple to apply and teach. It satisfies both the strong and the weak economic criteria.

If the criteria developed in Chapter 2 are taken as suitable criteria for selecting a method of audit sampling then the MUS method appears to satisfy these criteria more effectively than the alternative methods reviewed.

If the error <u>rate</u> is high, say above 10%, ratio estimate with stratification may prove the more efficient method.

8.21 <u>Conclusions on SS methods best suited to external auditing.</u>

In our opinion acceptance sampling is the sampling method best suited to testing accounting populations for degree of compliance with accounting procedures.

We consider the MUS approach best suited to testing accounting populations when the objective of the test is to estimate an upper bound on <u>error value</u>.

NOTES ON CHAPTER 8

1. Traditional audit samples would have to be doubled or even trebled to provide this degree of accuracy.

2. Personal communication from Robert Elliott, Partner, PMM, New York.

3. Personal discussion between author and Mr. A.C. Dekkers and Professor J. Krienz. Rotterdam June 26, 1980. See Krienz and Dekkers (1980)

4. See Chapter 3 for a fuller discussion of this point.

5. This comment ignores the possibility of using MUS for compliance testing. See Chapter 8 for discussion of limitations on this approach.

6. Attributes sampling can be used for substantive testing where the audit units are physical units not value items.

7. The SD could be estimated from the same accounting population in previous years.

8. The author found this operation to be a particularly tedious one. Have any of the advocates of this method ever tried it out in practice?

9. PMM, New York, Computer Audit Package

10. So long as the correlation between recorded and audited values is greater than 50%

11. Professor John McCray presented his method to an audit symposium organised by DHS, 1980. See McCray (1980)

12. MUS usually assumes the worst possible distribution of error value.

12(a). Comment by reviewer.

"Statement No. 3 is only correct in an audit sense as one may evaluate a p.p.s. sample in a classical sense to get a point estimate and a standard error. The procedure is in the Auditape and was used extensively some years ago by the Internal Revenue Service to measure conformance to Price Commission requirements."

13. By true value is meant post audit recorded value if every unit were audited as the audit sample is audited.

14. Unfortunately this method is tedious for large samples without a computer and 'adding up' may be quicker and cheaper.

15. Assuming low error rates less than, say, 5%. This is almost always overstatement error but understatement error is possible using the DUS version of MUS.

16. This is only a control figure for alpha risk, it does not mean we expect 1% of recorded value to be in error.

17. One firm (DHS) provide sampling tables with a beta risk as high as 61%! See DHS "Audit Sampling" 1979 page 57.

18. Suppose that materiality is taken as 10% of a gross of tax profit say, 20% on invested capital. Materiality is thus 2% of invested capital. Materiality is thus likely to be more than 2% of any individual account audited.

19. Private conversation with author. Toronto, Canada, July 1979.

20. Assuming that error value is a positive proportionate function of the value of the unit containing the error. The analysis in Chapter 2 suggests that it is.

21. McCray reports that using the Stringer bound the maximum beta risk can be several times the nominal specified beta risk under certain error conditions. Personal communication to TWM dated October 30th 1980.

22. Some comprehensive hypergeometric tables have been prepared. They have the advantage of calculating probabilities for smaller populations. The Poisson tables usually assume an infinite population.

Chapter 9.

A suggested scheme for selecting a statistical sampling method best suited to external audit.

9.1. Introduction

The previous chapters have examined several scientific sampling plans which, it has been claimed, are suited to audit work.

In this chapter we will suggest a strategy for selecting that sampling plan best suited to external audit.

The ideal solution would be a single sampling plan suited to all types of audits and auditors. However, we will argue that, at present, this is too ambitious a target. A more realistic objective might be to provide guidance to an auditor so that he can choose that sampling plan, from among several plans, which best satisfies the specific conditions of the audit on which he is engaged.

9.2. Objectives of audit sampling.

As noted in Chapter two above there are two broad objectives in external audit sampling. First the auditor wishes to test compliance with accounting procedures, second he wishes to place an upper limit on error value.

If his aim is to test for compliance with given accounting procedures then the audit sampling objective is to derive a sufficient assurance from the sample that the population under audit does not contain more than a given proportion of errors (1).

If the aim of the audit is to make a substantive test of audit values, then the audit sampling objective is to derive a sufficient assurance from the audit sample that the error value remaining in the population after audit does not exceed a given material value.

The 'materiality' value is likely to be derived independently of the particular population under audit. Thus to derive an audit sample objective in quantatitive terms, the auditor must decide on a minimum unacceptable upper limit on error rate or a minimum unacceptable upper error limit (UEL) on error value.

The latter will be some fraction of "materiality". The UEL is set lower than materiality to allow for alpha risk (see chapter eight).

The auditor must also decide on the level of reliability he requires in his inference from the audit sample. This level of reliability will be determined by his prior knowledge of the population under audit, knowledge derived from analytical review, the previous year's audit papers, similar audits, internal audit reports and such like.

The reliability so derived is reliability as to beta risk. The risk of accepting a population that should be rejected. This is the primary risk in auditing. But the auditor must also build in some allowance for alpha risk. The risk of rejecting an acceptable population.

We consider that achieving an acceptable alpha risk is a constraint on, rather than an objective of, selecting a suitable sampling plan, so we will discuss the matter further in the following section.

The objectives of audit sampling can thus be represented by two factors both determined subjectively by the auditor, namely,

(a) Reliability

(b) Minimum unacceptable upper limit on the estimate of the error rate or error value.

9.3. Constraints on selecting an external audit sampling plan.

In chapter two we identified certain desirable characteristics of an audit sample and an audit sampling process.

The audit sampling process is also subject to certain constraints over which the auditor may have no control. These constraints are derived from four sources namely:

(a) The statistical characteristics of the accounting population under audit.

(b) The characteristics of the auditor.

(c) The medium on which the accounting population under audit is stored.

(d) The economics of the audit process.

A sampling method suited to external audit work must attempt to satisfy as many of these constraints as possible before it can be considered satisfactory as an audit sampling method.

We will now examine these various constraints in rather more detail.

9.3.1. The statistical characteristics of accounting populations.

As noted in chapter two accounting populations tend to be highly skew as to value. This sets up a high coefficient of variation which in turn generates a high sample size unless some form of stratification is used.

As noted in chapter three error value is also highly skew in most accounting populations therefore we must beware against assuming that the central limit theorem can be applied to inferences derived from samples of error value(2).

Other statistical characteristics which may impose a constraint on the sampling method chosen are:

(a) the rarity of error.

(b) the existence of zero and negative balance.

(c) the existence of understatement of recorded amounts which could be, theoretically, of infinite value.

(c) the characteristic of 'tainting' i.e. that an amount in error is unlikely to be 100% in error but rather tainted by a given fraction of 100%.

A sampling method aimed at estimating an upper limit on the amount of monetary error should be designed to handle as many of these constraints as possible.

9.3.2. <u>The characteristics of the auditor.</u>

Few auditors possess much depth of knowledge in statistical theory. An ideal audit sampling method should be simple both to apply and to understand. The simplicity in application may not be so important if a computer is available. In this case the preparation of the documentary input to the computer and the presentation of the sampling results should be simple to understand.

If a sampling plan can be applied both manually and by computer this must be considered a strong point in its favour.

A sampling plan may be simple to apply but difficult to understand.

A complicated sampling plan need not impose an absolute veto on its use but clearly if the logic behind a sampling plan can be understood easily, it must be viewed favourably compared to a plan derived via a complex logical process.

An ideal audit sampling plan should be simple enough for an auditor to apply it without undue difficulty and understand it without undue effort.

Much depends upon the degree of delegation employed by the audit firm employing the system. If the audit firm employs a highly structured, highly centralised audit procedure using staff of limited knowledge and ability then the audit sampling system must be simple to apply.

9.3.3. <u>The storage medium</u>.

If the accounting population under audit is stored on a computer it will be possible to employ a more sophisticated sampling plan than if it is not.

The computer can provide the following services at relatively low cost. The computer can,

(a) generate sets of random numbers, (as in cell sampling).

(b) calculate sample size and 'skip interval' given the required reliability and precision.

(c) Stratify the population in accordance with prescribed conditions.

(d) select the audit sample under even very complicated sample selection rules, as in stratified mean-per-unit sampling.

(e) Evaluate the results of the audit sample test.

(f) Decide on the next audit step to take if the results of the sample test are unsatisfactory.

All of these procedures can be delegated to the computer if the accounting population is stored on the computer and an SS audit programme is available. Needless to say every one of these steps _need_ not be so delegated. The auditor will decide which steps he prefers to delegate to the computer. The absence of a computer is a powerful constraint on the choice of sampling plan in audit work.

Several audit sampling techniques can only be used if the accounting population under audit is stored on a computer system. For example, stratification procedures are only feasible if the population to be stratified is stored on a computer, and cell sampling is cumbersome without a computer.

The absence of a computer, therefore, reduces the number of sampling options which can be considered.

9.3.4. _The economics of the audit process_.

As noted in chapter seven the economics of the audit process impose a possible constraint on the choice of sampling method.

Audit sample sizes used in North North America appear, on the average, to be about double the audit sample sizes used in the United Kingdom (Section 6.7.1 and 6.7.2). An audit sampling method is of little use if it generates sample sizes far above those traditionally drawn by the auditor.

Objective

Test whether the total value of error in the population is likely to exceed materiality.

Constraints

Skewness of accounting population	A1
Skewness of error values	A2
Rarity of error	A3
Zero account balances	A4
Negative account balances	A5
Understatements	A6
Tainting i.e. partial errors	A7
Simplicity of application	S1
Simplicity of understanding	S2
Computer required	C1
Average size of sample $\leqslant$ traditional sample size	E1
Learning and set-up cost acceptable	E2
Alpha risk can be calculated	P1

Exhibit 9.1. Objective and constraints on sampling plan to test whether error value in population under audit exceeds a material amount

If the cost of learning and set up is unduly heavy this may also present an economic obstacle to adopting a particular system.

Finally, we must consider the problem of alpha risk. If the UEL is set equal to materiality the sampling method becomes a discovery sampling method. The discovery of a single error causes the population to be rejected.

Most MUS methods solve this problem by fixing £U the upper error limit well below materiality.

However, if many small value errors occur too many acceptable populations may still be rejected.

Practitioners of MUS tend to deny that this is a problem (3) but we believe that it <u>ought to be</u> a problem if the distributions of error values noted in chapter three are representative.

A sampling method designed to test for error value should, therefore, be able to handle alpha risk. The inability of a sampling method to specify alpha risk must be considered a limitation on that sampling method.

<u>A listing of constraints on value sampling</u>.

In exhibit 9.1 we have listed the various constraints which an ideal audit sampling method should satisfy. The audit sampling in this case is aimed at substantive testing of monetary value.

In a later section of this chapter we will examine each value sampling method to see how well they can satisfy each of these constraints.

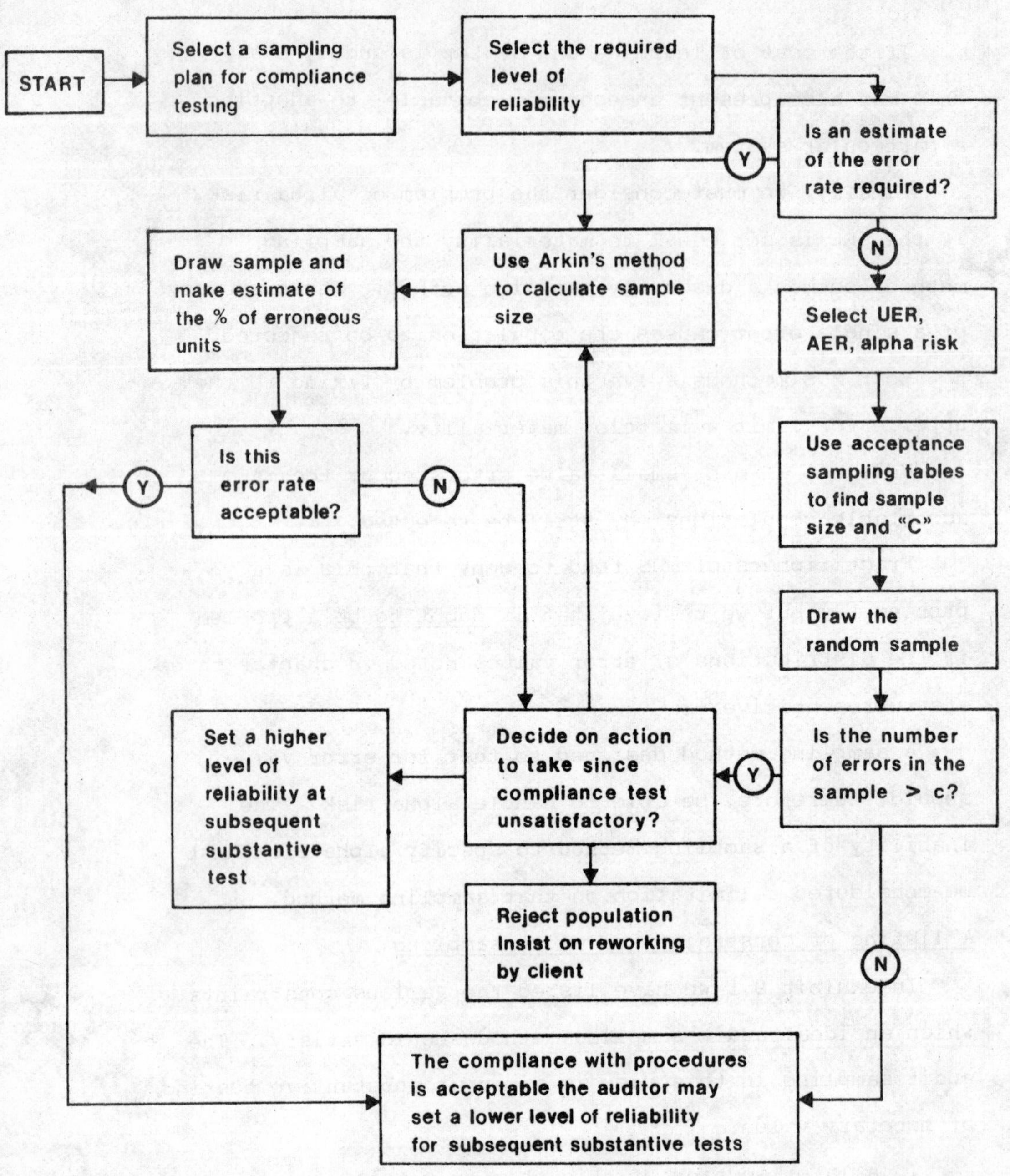

Exhibit 9.2. A flow diagram to assist in selecting a suitable sampling plan for compliance testing.

9.4. Selecting a sampling plan for compliance testing.

We will now suggest a strategy for selecting that audit sampling plan best suited to compliance testing of procedures and substantive testing of values.

First we will describe our strategy for selecting a suitable sampling plan. Later we will comment on the logic behind our selection process.

In chapter eight we argued in favour of using acceptance sampling for compliance testing. The flow diagram presented in exhibit 9.2 explains how we arrived at this conclusion.

(a) The first step is to decide on the level of reliability required in the inference from the audit sample. The level of reliability required will, in the case of procedural testing, depend upon the consequences flowing from the decision based on the inference.

(b) The auditor must decide whether an estimate of the true error rate is required. Under normal circumstances his requirement will be less stringent. The auditor need only assure himself that the true error rate is likely to be less than some upper bound.

If an estimate is not required we suggest that the auditor use acceptance sampling for the reasons presented in chapter eight. The auditor will thus adopt an hypothesis testing approach. If an estimate is required we suggest that the auditor use Arkin's "confidence interval approach" described in chapter eight, and in Arkin (1976).

(c) The auditor has decided to use acceptance sampling. He has already decided on his acceptable beta risk so he must further decide on his minimum unacceptable error rate (UER), acceptable error rate (AER) and alpha risk. Alpha risk can be high, say 20%, since the consequences of incorrect rejection will normally be no more than an increase in the required reliability at the subsequent substantive test.

(d) Use acceptance sample tables to calculate the sample size and the critical number of errors.

(e) Draw the sample, audit the sample units, and calculate the number of compliance errors "c".

(f) If the errors do not exceed the critical number, accept the population as being of acceptable quality.

If errors exceed the critical number decide on further action.

(g) Errors exceeded critical number, therefore the true error rate is probably unacceptable.

The auditor must decide on further action such as:

(h) Hand population back to client for reworking.

(i) Modify reliability required for subsequent substantive testing of values.

(j) Estimate true error rate.

If the critical number is reached quickly the auditor may be tempted to reject the population outright and demand a reworking by his client. Alternatively he may decide to estimate the true error rate before proceeding to such a drastic step.

Under normal circumstances the consequence of rejection will be that the auditor adjusts his reliability level at the subsequent substantive testing stage of the audit.

(j) The auditor may decide that a hypothesis testing approach is unsuitable and that an estimate of the true error rate is needed. The possible reasons for this decision are discussed in chapter eight.

We suggest that Arkin's method called the "confidence interval approach" should be used. This method supplies an estimate if the error rate lies within the critical region between UER and AER. (See Arkin 1976).

(k) Draw a sample and make an estimate of the actual error rate.

(l) Decide whether the error rate is low enough to suggest that the population under audit is of acceptable quality.

(m) Accept the population under audit as being of acceptable quality. The consequence flowing from this decision will normally be to accept a relatively high beta risk at the later stage of substantive testing of value.

If the estimated error rate is not acceptable the auditor is faced with the same range of choices listed under (g). However, it can be argued that an estimate of the true error rate may provide a better decision.

Exhibit 9.2. sets out our suggested strategy for choosing a sampling method suited to compliance testing of accounting procedures.

9.4.1. Commentary on compliance testing strategy.

In chapter eight we presented our reasons for preferring the hypothesis testing approach of acceptance sampling to the survey sampling approach.

The main arguments against the survey sampling approach are that (a) an estimate of the true error rate is not needed if the error rate is likely to be below UER and (b) that traditional audit sample sizes in the UK would not be likely to provide an estimate of sufficient precision to be useful.

The main argument in favour of acceptance sampling is that it leaves less discretion to the junior auditor. The system can be programmed in advance by senior staff.

Several audit firms use the MUS method for compliance testing. It may be that the economics of audit sampling enforce this approach. Whatever the reason we are of the opinion that it is a second best solution.

We are not convinced that MUS is suited to compliance testing for the following reasons.

(a) The audit sample is biased towards units of large value. These units may not be representative of the population as a whole. Some empirical evidence i.e. Johnson, Leitch, Neter (1980) suggests that the error rate among high value units is higher than average. It is also intuitively appealing to suggest that in some audits the control of high value units may be much tighter than on low value units resulting in <u>lower</u> error rates rather than average, for high value units.

In such cases a PPS sample will give us a biased estimate of the population error rate.

(b) Compliance testing should logically precede substantive testing since the compliance test will contribute towards deciding the reliability desired on the inference from the

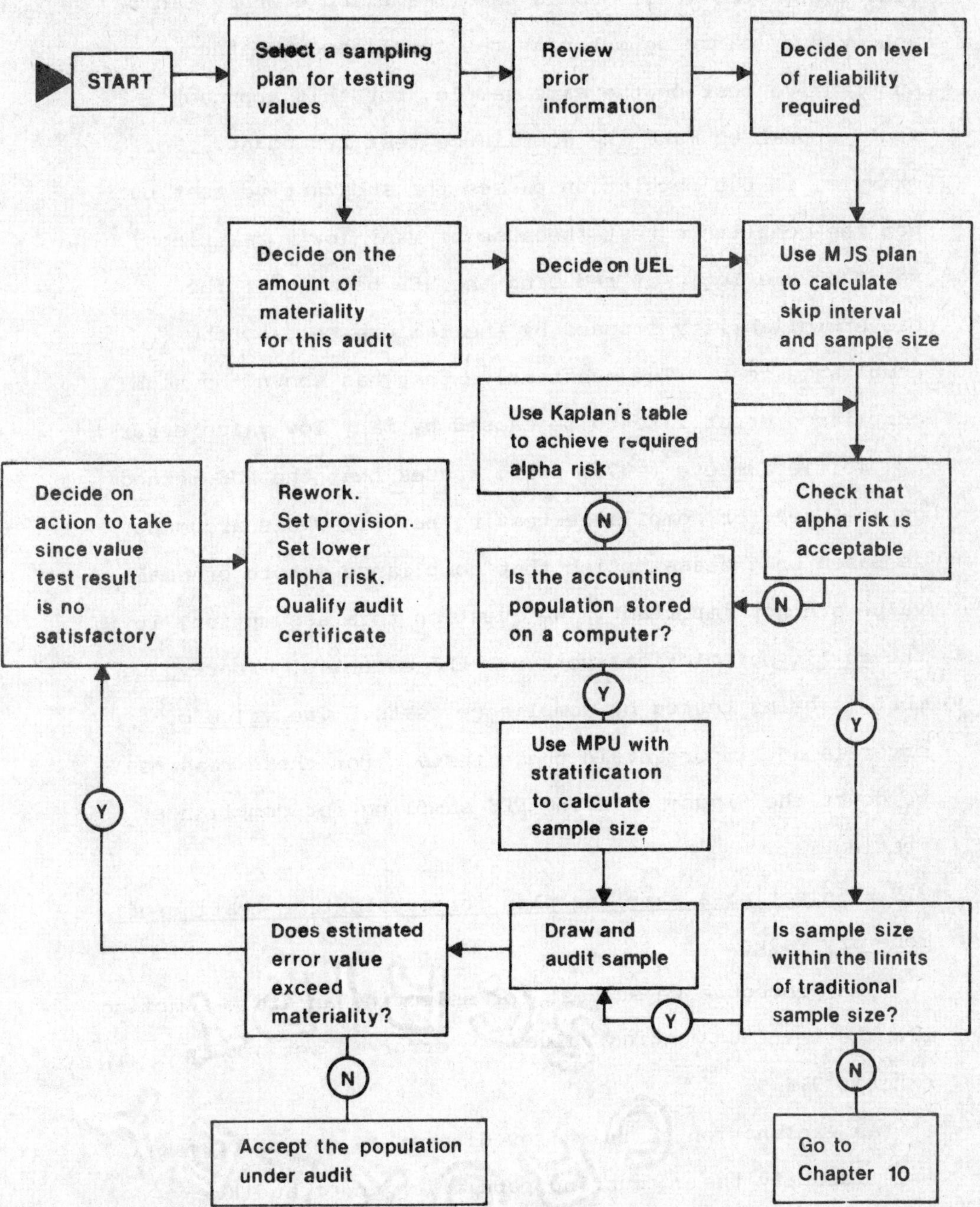

Exhibit 9.3. A flow diagram to assist in selecting a suitable sampling plan for testing accounting values.

substantive test. If MUS is used, sampling economy can be achieved by carrying out both the compliance and the substantive test on the same sample, but this approach would appear to make the compliance test redundant. For example, if the population passes the substantive test but not the compliance test (because of many low % taintings) we doubt the logic of reducing the UEL because of the higher reliability imposed by the failure to pass the compliance test. The substantive test has shown the high compliance error rate to be caused by many low value errors!

Teitlebaum etc (1975) p.11, argues that the DUS method can be used for compliance testing but his whole argument is based on the assumption that compliance errors of small value are not important. We question this assumption, it is the quality of clerical work not the extent of error value that is being tested by compliance tests. The value of error is not important in these tests. For these reasons we doubt the wisdom of using PPS sampling for compliance testing.

9.5. Selecting a sampling plan for substantive testing of monetary value.

The procedure we suggest for selecting an audit sampling plan to test accounting values error is set out in exhibit 9.3.

An explanation of this flow diagram is as follows:

(a) Identify the accounting population under audit.

(b) Review prior information about the population. This will include analytical review, previous years audit papers, internal audit reports and the results of prior compliance tests.

(c) Use the information collected under (b) to decide on the required level of reliability (beta risk). It is assumed that total assurance required from all sources will be the same for all populations under audit.

(d) Find the level of materiality (£M) agreed upon for this audit. This amount will have been determined independently of the particular population under audit.

(e) Use this materiality amount to decide on the acceptable upper error limit (£U). Many auditors appear to fix the UEL at 50% of materiality. The ratio U/L will be affected by the number of errors expected in the population under audit. (See Teitlebaum etc (1975) p.31).

(f) Use the MUS method to calculate the required sample size. In particular take the reliability level decided under (c) and the UEL decided under (e) to calculate the "skip interval", £S., and audit sample size. An incremental Poisson table such as that illustrated in exhibit 9.4. will assist in making this calculation.

(g) Check the sampling plan for alpha risk by using Kaplan's table (Roberts 1978, p.253) or the formula provided therein.

(h) If alpha risk is too high because the proportion of expected errors is too high check whether the accounting population is stored on a computer.

(i) If it is <u>not</u> then recompute sample size/skip interval for maximum acceptable alpha risk. We suggest a maximum alpha risk level of 20%. The acceptable alpha risk will depend on

UEL TABLE

(for calculating upper error limits of either overstatements or understatements)

Confidence levels								Number of sample errors	Confidence levels					
75%		80%		85%		90%			95%		97.5%		99%	
UEL	PGW	UEL	PGW	UEL	PGW	UEL	PGW	errors	UEL	PGW	UEL	PGW	UEL	PGW
1.39	—	1.61	—	1.90	—	2.31	—	0 (BP)	3.00	—	3.69	—	4.61	—
2.70	.31	3.00	.39	3.38	.48	3.89	.58	1	4.75	.75	5.58	.89	6.64	1.03
3.93	.23	4.28	.28	4.73	.35	5.33	.44	2	6.30	.55	7.23	.65	8.41	.77
5.11	.18	5.52	.24	6.02	.29	6.69	.36	3	7.76	.46	8.77	.54	10.05	.64
6.28	.17	6.73	.21	7.27	.25	8.00	.31	4	9.16	.40	10.25	.48	11.61	.56
7.43	.15	7.91	.18	8.50	.23	9.28	.28	5	10.52	.36	11.67	.42	13.11	.50
8.56	.13	9.08	.17	9.71	.21	10.54	.26	6	11.85	.33	13.06	.39	14.58	.47
9.69	.13	10.24	.16	10.90	.19	11.78	.24	7	13.15	.30	14.43	.37	16.00	.42
10.81	.12	11.38	.14	12.08	.18	13.00	.22	8	14.44	.29	15.77	.34	17.41	.41
11.92	.11	12.52	.14	13.25	.17	14.21	.21	9	15.71	.27	17.09	.32	18.79	.38
13.03	.11	13.66	.14	14.42	.17	15.41	.20	10	16.97	.26	18.40	.31	20.15	.36
14.13	.10	14.78	.12	15.57	.15	16.60	.19	11	18.21	.24	19.68	.28	21.49	.34
15.22	.09	15.90	.12	16.72	.15	17.79	.19	12	19.45	.24	20.97	.29	22.83	.34
16.31	.09	17.02	.12	17.86	.14	18.96	.17	13	20.67	.22	22.24	.27	24.14	.31
17.40	.09	18.13	.11	19.00	.14	20.13	.17	14	21.89	.22	23.49	.25	25.45	.31
18.49	.09	19.24	.11	20.13	.13	21.30	.17	15	23.10	.21	24.75	.26	26.75	.30
19.57	.08	20.34	.10	21.26	.13	22.46	.16	16	24.31	.21	25.99	.24	28.04	.29
20.65	.08	21.44	.10	22.39	13	23.61	.15	17	25.50	.19	27.22	.23	29.31	.27
21.73	.08	22.54	.10	23.51	.12	24.76	.15	18	26.70	.20	28.45	.23	30.59	.28
22.81	.08	23.64	.10	24.63	.12	25.91	.15	19	27.88	.18	29.68	.23	31.85	.26
23.89	.08	24.73	.09	25.74	11	27.05	.14	20	29.07	.19	30.89	.21	33.11	.26
24.96	.07	25.82	.09	26.85	.11	28.19	.14	21	30.25	.18	32.11	.22	34.36	.25
26.03	.07	26.91	.09	27.96	.11	29.32	.13	22	31.42	.17	33.31	.20	35.61	.25
27.10	.07	28.00	.09	29.07	.11	30.46	.14	23	32.59	.17	34.52	.21	36.85	.24
28.17	.07	29.09	.09	30.18	.11	31.59	.13	24	33.76	.17	35.72	.20	38.08	.23
29.24	.07	30.17	.08	31.28	.10	32.72	.13	25	34.92	.16	36.91	.19	39.31	.23
30.31	.07	31.25	.08	32.38	.10	33.84	.14	26	36.08	.16	38.10	.19	40.54	.23
31.38	.07	32.33	.08	33.48	.10	34.96	.12	27	37.24	.16	39.29	.19	41.76	.22
32.44	.06	33.41	.08	34.58	.10	36.08	.12	28	38.39	.15	40.47	.18	42.98	.22
33.50	.06	34.49	.08	35.68	.10	37.20	.12	29	39.55	.16	41.65	.18	44.19	.21
34.56	.06	35.56	.07	36.77	.09	38.32	.12	30	40.70	.15	42.83	.18	45.41	.22
39.86	.06	40.93	.08	42.22	.09	43.88	.12	35	46.41	.15	48.68	.17	51.41	.20
45.14	.06	46.27	.07	47.64	.09	49.39	.10	40	52.07	.13	54.47	.16	57.35	.19
50.40	.05	51.59	.06	53.04	.08	54.89	.10	45	57.70	.12	60.21	.15	63.24	.18
55.64	.05	56.90	.06	58.40	.07	60.34	.09	50	63.29	.12	65.92	.14	69.07	.17
60.87	.05	62.18	.06	63.76	.07	65.79	.09	55	68.86	.12	71.61	.14	74.87	.16
66.09	.05	67.46	.06	69.10	.07	71.20	.08	60	74.40	.11	77.22	.13	80.60	.15
71.30	.04	72.73	.05	74.43	.07	76.61	.08	65	79.91	.10	82.85	.12	86.36	.15
76.50	.04	77.98	.05	79.74	.06	81.99	.08	70	85.41	.10	88.45	.12	92.09	.14
81.70	.04	83.23	.05	85.05	.06	87.37	.08	75	90.89	.10	94.02	.12	97.74	.14
86.90	.04	88.46	.05	90.34	.06	92.73	.07	80	96.36	.09	99.58	.11	103.39	.13
92.08	.04	93.69	.05	95.62	.06	98.09	.07	85	101.80	.09	105.11	.11	109.04	.13
97.26	.04	98.92	.04	100.90	.05	103.42	.07	90	107.24	.09	110.61	.10	114.69	.12
102.43	.03	104.14	.04	106.17	.05	108.76	.06	95	112.67	.08	116.15	.10	120.26	.12
107.60	.03	109.35	.04	111.43	.05	114.07	.06	100	118.07	.08	121.61	.09	125.84	.11

* Computed from the cumulative Poisson distribution

Exhibit 9.4. An incremental Poisson Table for calculating sample size in DUS sampling, (from Leslie etc 1980)

the cost of overauditing.

(j) If the accounting population is stored on a computer, consider the possibility of switching to mean-per-unit sampling with stratification. This course is only viable if a good number of value errors are expected in the population under audit, say in excess of 5%.

This plan may be best for auditing inventory where many small errors are likely to occur.

(k) Is the audit sample size generated by the previous calculations within the size traditionally drawn in the UK by external auditors?

(l) If it is <u>not</u> we have a problem. This problem will be discussed in the next chapter.

(m) If the sample size is within acceptable limits the auditor proceeds to audit the sample in the usual way and list all errors discovered. In particular the tainting % of each error is noted and listed ordinarily from largest to smallest.

(n) The upper bound on error value is calculated. Several alogorithms are available for calculating this bound. For example.

The Stringer bound (DHS 1979)

The cell bound (Leslie etc 1980)

The CAV bound (Neter and Loebbecke 1975)

The McCRAY bound (McCray 1980)

We prefer the cell bound since it has been widely used and is simple to apply and understand.

The McCRAY bound is a promising alternative but insufficient evidence of its successful application in practice is available at the time of writing.

Once the upper bound on the estimate of error value is calculated the upper bound is compared to the UEL calculated under (e) above.

(o) If the estimated upper bound on the error value exceeds materiality the auditor has several courses open to him.

(1) He can reduce his alpha risk further by increasing his sample size. This course will only be chosen if drawing an increased sample size is practical and not too expensive. The auditor then proceeds for a second cycle through the flow diagram.

(2) He can hand the population under audit back to the client for reworking.

(3) He can set up a provision equal to the excess of estimated error value over acceptable error value. (See Teitlebaum etc 1975 p.16).

(4) He can qualify his audit certificate.

(p) If the upper bound on the estimate of error value is below the UEL calculated in (e) above, the auditor will accept the population under audit as being of acceptable quality.

9.5.1. <u>Commentary on substantive testing strategy</u>.

In chapter eight we argued the case for the MUS method being superior to other audit sampling methods as a <u>general</u> method. Exhibit 9.1. listed the various constraints which should be satisfied by a statistical sampling method suited

Sampling method.	Constraints not handled satisfactorily by given sampling method.
Mean per unit	A1 A2 A6 E1
Mean per unit (with stratification)	A2 A3 A6 S1 S2 C1 E2
Difference estimation	A1 A2 A3 A6 E1
Difference estimation (with stratification)	A2 A3 A6 S1 S2 C1 E2
Ratio estimation	A1 A2 A3 A6 E1
Ratio estimation (with stratification)	A2 A3 A6 S1 S2 C1 E2
MUS methods	A4 A6 P1

Exhibit (9.5) List of constraints which are _not_ well satisfied by various statistical sampling methods when applied to audit work. See Exhibit (9.1) for coding of constraints.

Note: All constraints are not of equal weight. The sampling plan which we have chosen was selected because of the weights which we allocated intuitively to the particular constraints. Another auditor may allocate a different set of weights and so choose a different plan.

to external auditing of value. Exhibit 9.5 lists those constraints which <u>cannot</u> be adequately satisfied by each sampling method. Note that the MUS method can satisfy more constraints than any other SS method.

Ths MUS method handles the problem of skewness very neatly, it is simple to operate and economical in sample size.

The limitations on the MUS method are concerned with zero value balances, understatements and specification of alpha risk. As we explained in chapter eight all of these limitations can be overcome at some cost in additional complexity.

If many errors of low tainting occur it may be prudent to use mean-per-unit with stratification. But this latter option can only be adopted if the accounting population under audit is stored on a computer.

If the audit sample size generated is much larger than the size traditionally drawn in audit sampling, the auditor has a problem. He has various options, he can either reduce reliability, increase UEL or even decide not to use statistical evaluation methods. We shall discuss these options further in the next chapter.

If the estimated upper bound on error value exceeds the UEL the situation facing the auditor is identical to that facing him if a traditional audit sample contains an unacceptable number of errors. He can increase his sample size, have the population rechecked, set up a provision, or, as a last resort, qualify the audit certificate.

9.6. Summary and conclusions.

In selecting a statistical sampling method suited to external auditing the auditor must first agree on the precise objective of the audit and the various constraints which stand in the way of achieving this objective.

In compliance testing the objective is usually defined in the context of a given level of reliability and an acceptable and unacceptable error rate.

In substantive testing of monetary values the objective is usually defined in the context of a given level of reliability, derived from prior knowledge, and materiality.

There are many constraints on efficient audit sampling and the auditor must identify these before he selects an efficient audit sampling plan.

We present a flow diagram describing a suggested strategy for selecting a suitable audit sampling plan, both for compliance testing of procedures and for substantive testing of monetary values.

In general we favour acceptance sampling for compliance testing of procedures and some variant of MUS for substantive testing of values.

There are special circumstances in which other sampling methods should be used.

Notes on Chapter nine.

1. The auditor would also be influenced by the quality of the errors he discovered.
2. As assumed, for example, by Reneau (1978).
3. Don Leslie of CG, an experienced pracitioner of the DUS method, holds this opinion. (Personal conversation with D.L.).

Chapter 10.

The problem of excess sample size.

10.1. The problem of excess sample size.

The use of statistical sampling by the external auditor need not result in an increase in average audit sample size. The only study to examine this question, Aly and Duboff (1971) concluded that SS reduced audit sampling size compared to the size of traditional audit samples.

If, however, SS should increase average audit sample size significantly above the size used in traditional audit sampling the auditor is faced with a problem.

Exhibit 9.3. in the previous Chapter presented a flow diagram of the logic behind a suggested scheme for selecting a sampling plan for testing accounting values.

This diagram included one uncompleted branch which consisted of the question "what will the auditor do if the audit sample size exceeds a size which he considers to be economically viable"?

Audit sample sizes appear to be determined by tradition rather than by logic. Traditional sampling methods do not appear to use any formal system to determine sample size.

If the sample size determined by formal statistical methods exceeds the traditional sample size by a significant margin the auditor is faced with a difficult problem. What should he do?

This chapter of the study is devoted to a discussion of possible solutions to this problem.

Let us be clear on the precise situation facing the auditor. (a) He has determined materiality and required reliability of the estimate from prior knowledge of the accounts. (b) He has also decided on a sampling method suited to auditing this type of accounting population.

(c) He uses the required reliability and precision to calculate audit sample size. (d) He finds that the sample size so determined significantly exceeds the traditional sample size for this type of audit. Thus if he draws this size of sample the cost of the audit will be significantly increased.

10.2. Options available.

What options are open to the auditor under these conditions? We will examine the following options.

(a) Increase the audit fee.

(b) Discontinue the use of statistical sampling.

(c) Increase materiality or reduce the required reliability of the estimate (or both).

(d) Devise a more efficient sampling plan.

(e) Work with the internal auditor to devise integrated samples.

Let us examine each of these options in turn.

10.2.1. Increase the audit fee.

Comment

This is the simplest course to follow but it rather begs the question.

It may be that a significant increase in sample size will not generate a significant increase in audit fee.(See Chapter 7) Alternatively it may be that the client will consent to an increase in fee to ensure a proper audit, when the situation is explained to him.

We suspect that both of these possibilities are exceptional and that under normal circumstances an increase in audit fee will meet resistance from the client.

Auditors are under great pressure at present to keep down audit fees (1) and the size of audit fees are currently (1981) the subject of negative criticism in the UK press.

We conclude that this option, an increase in fee, will not be available in most cases.

10.2.2. Discontinue the use of statistical sampling.

Comment.

Since many professional accounting firms, including several major firms, do not use SS in their auditing procedures the auditor who faces an excessive SS sample might be tempted to discontinue using the method.

We trust that the auditor will resist this temptation.

The auditor has defined his precise requirements from the audit sample, he has then calculated the minimum sample size needed to meet these requirements, but, finding this audit sample too large he ignores the statistical approach and switches back to non-statistical sampling and takes a smaller sample. In effect he has changed his specifications by making them less rigerous, yet concealed the fact by ignoring statistical inference.

This option is surely illogical. An analogy might be if a man trapped in a cage with a tiger on a dark night believes he can solve his problem by switching out the light.

As we shall see in a moment, the auditor may well decide to alter his initial specifications but he must surely do this within the framework of SS not outside the framework.

Opting out of SS because the initial conditions set result in too large a sample is surely a defeatist council, not to be supported by those attempting to improve the efficiency of audit procedures.

10.2.3. Alter the required reliability and/or precision.

Comment

If the cost of sampling imposes an absolute constraint on increasing the audit sample size the auditor may decide to make reliability or precision the dependent variable (or both).

This alters the audit strategy suggested in chapter two.

The reader will recall that the strategy outlined in chapter two set the auditor the task of determing materiality and reliability prior to calculating audit sample size. Sample size was the dependent variable.

Under the new approach suggested above, sample size is fixed by the audit fee. The auditor must either reduce his reliability or increase his materiality to a level determined by the sample size (2).

Since materiality is determined independently of the specific population under audit we would recommend that the auditor reduces the level of reliability he requires in his inference.

This solution seems to follow the logic of the audit situation. The auditor achieves the highest degree of reliability possible given the level of audit fee. If the client wishes the auditor to achieve a higher level of reliability he can pay a higher level of audit fee.

The auditor is not a magician. He cannot be expected to squeeze a quart out of a pint pot.

10.2.4. Develop a more efficient sampling plan.

Although an auditor cannot be expected to squeeze a quarter out of a pint pot, he might manage to squeeze a few more gills. He might be able to improve his audit sampling methods so that more information can be squeezed out of a given sample size.

Alternatively he might conclude that existing SS methods are too conservative and that by squeezing out some of the excess conservatism he can reduce sample size.

Auditors and statisticians have put in a great deal of effort over the last twenty years in trying to improve the efficiency of audit sampling.

The application of PPS sampling and Bayesian statistics to audit work are examples of this improved methodology. Kinney (1975), has recently suggested an interesting new line of research by linking regression analysis at the analytical review stage of the audit to determining audit sample size at the subsequent sampling stage.

Other interesting possibilities were described in chapter five and eight. For example, current research is concentrating on squeezing the excess conservative element out of MUS sampling.

Most MUS upper error limits are calculated on the assumption that all undiscovered errors are 100% tainted. Chapter three suggests that this is a very conservative assumption. Since most discovered errors are tainted by much less than 50% it is most unlikely that all undiscovered errors are 100% tainted,

Mottershead (1980) has suggested that undiscovered errors should be assumed to be only 50% tainted. This assumption would result in a marked reduction in sample size for any given level of materiality.

McCray (1980) has put much effort into devising methods to reduce audit sample size without unduly complicating the audit process. He has introduced a sample value correction factor to allow for the frequency of errors occurring in amounts of different value and devised tables to allow for different patterns of error tainting.

Leslie etc (1980) introduce a number of ideas to reduce audit sample size in their book on DUS.

Progress in reducing audit sample size in SS has been very slow but sure. The main obstacle is the sluggish diffusion of these more efficient sampling methods within the audit profession.

If an auditor decides that he can afford to spend £x on audit sampling it behoves him to ensure that he extracts the maximum possible information from the £x.

As a professional auditor he ought to know the most efficient sampling method available.

It seems likely that in the not too distant future these developments in sampling theory together with empirical investigations of accounting error will help to reduce audit sample size at any given level of assurance.

If the sample size suggested by SS is in excess of the sample size suggested by traditional auditing the auditor should study these recent developments in SS to see whether he considers that they allow him to reduce his sample size further.

10.2.5. <u>Work with the internal auditor to devise integrated samples</u>.

The external auditor is not the only auditor in many firms. Many UK companies employ a team of internal auditors to perform a continuous audit of the accounting system.

The internal auditor also uses sampling methods to test the accounting system under audit. The total of the samples drawn by the internal auditor in any one year is likely to be far in excess of the size of sample drawn by the external auditor. Small internal audit samples are drawn at regular intervals during the year but build up into samples of substantial size by the year end.

If the external auditor can persuade the internal to work with him in developing an integrated sampling strategy, most of the problems discussed in this chapter can be solved. The total sample, internal plus external, is likely to exceed the size at which the usual levels of materiality (5-10% of profit) and reliability (80% - 95%) can be catered for. (3).

There is no legal impediment preventing an external auditor from using data supplied by an internal auditor although the external auditor cannot slough off any of his responsibility onto the internal auditor.

Several internal auditors declared to the author that they liaise with the external auditor on the overall audit of their firm. (4). However we are not aware of any instance of integrated audit sampling.

The integration of audit samples of flows such as sales or purchases is easily effected, so long as both auditors use the same sampling plan.

The audit of stocks (inventory, debtors, creditors, small tools etc) present a more difficult problem. These are audited at one point in time, not throughout the year. Therefore, regular internal audit samples do not build up as they do with flows.

Various solutions suggest themselves. The internal auditor could be persuaded to draw a larger sample on the date of the external audit of the given account. The analysis of variance could be used to test whether the external audit sample findings conform to the internal audit sample findings drawn during the year, a regression chart could be drawn up to perform a similar test (5).

Whichever method is adopted the larger internal audit samples can be used to reduce the audit sample size, drawn by the external auditor. (1) As with DHS's STAR system.

10.3. <u>Summary and conclusions</u>.

The audit sample sizes generated by SS need not exceed the sizes generated by traditional audit method. If they do exceed traditional sample size by a significant margin several options are available to the auditor.

The auditor can discuss the matter with his client and try to persuade him of the need to pay for a satisfactory level of assurance. This course is likely to meet with resistance in the current climate of opposition to increased audit fees.

The auditor can discontinue the use of statistical evaluation of his assurance in his audit conclusions. This is surely a defeatist reaction not to be recommended.

The auditor can turn reliability into a dependent variable determined by the level of audit fee. The client, in other words, gets what he is prepared to pay for. This solution would seem to invert traditional audit philosophy.

The auditor can seek out a more efficient sampling plan. Much research is currently in progress with this objective in view. It is possible that current methods of MUS are too conservative in calculating an upper bound on error value.

Finally, the external auditor may attempt to persuade the internal auditor to co-operate in devising integrated audit samples to improve the confidence of the external auditor in his sample inferences.

Notes.

1. This point was emphasised many times in our discussion with partners in professional accounting firms.
2. Or play off one against the other.
3. Using DUS, CMA or MEST systems of sampling, a sample size exceeding 300 will usually satisfy these conditions.
4. 4 out of 32 internal auditors admitted to this practice during a seminar in internal audit conducted at the City University Business School 4th June, 1980.
5. This last approach is similar in technique to the STAR system employed by DHS.

Part B.

A brief study of some other aspects of statistical sampling.

11. Organisation.
12. Training.
13. The computer.
14. Legal implications.
15. Literature.

Chapter 11.

The organisational framework for using statistical sampling in external auditing.

11. Introduction.

How is statistical sampling to be integrated into the system of external audit?

We discussed this question at some length with practitioners of SS on both sides of the Atlantic.

From these discussions we conclude that a firm wishing to adopt SS should proceed on the following lines.

1. Make a series of policy decisions on SS.
2. Appoint a project co-ordinator.
3. Ensure access to a statistical expert.
4. Devise appropriate documentation.
5. Ensure that a suitable teaching programme is available.
6. Acquire an SS computer package.

Once these steps have been taken the organisational framework for using SS exists, but practitioners emphasised to us the fact that it takes several years of practice before an audit firm can use SS efficiently.

We will now examine each stage of the implementation process in more detail.

11.1. Policy decisions.

A policy decision to use SS must be taken at the highest level. The audit firm must adopt SS not an individual partner or manager.

This does not mean that a policy decision must be taken to use SS on all audits. The decision will be that SS can be used as an audit tool on appropriate audits.

In the first instance the decision should relate to a fixed number of years, after which the results of the experiment will be reviewed. The experimental period should be for at least two and preferably three years.

Once the decision has been taken to use SS as an audit tool a further decision needs to be taken as to which method of SS to use. Firms with no previous experience may prefer to start by using some form of acceptance sampling to estimate an upper bound on error rate. However one school of thought argues that audit firms thinking about adopting SS ought to move straight to monetary unit sampling which will provide an upper bound on error value at little additional cost compared to acceptance sampling.

Once it has been decided to use SS and the given method or methods have been chosen it still remains to set up appropriate criteria for deciding whether or not SS should be used on a particular audit.

The criteria for using SS may be set out as "guidelines" or as mandatory rules. Whether they are guidelines or rules depends upon the degree of control over audit method imposed at the centre by individual firms.

Our survey noted a wide variation between firms on this point.

11.2. <u>Appoint a project co-ordinator</u>.

Every firm in our survey who had adopted SS had appointed a member of staff to co-ordinate the various activities involved in using SS. This was not a full time activity in most cases.

The SS co-ordinator interviewed by us almost invariably had a mathematical, engineering or statistical background, but this, in our opinion, is not essential since a deep knowledge of statistics or mathematics is not needed to operate SS. Practical experience is, however, essential.

We asked these SS co-ordinators the kind of problems which were referred to them and, without exception, they stated that the problems referred were auditing problems rather than statistical problems.

An ideal project co-ordinator would be trained by an

international auditing firm who use SS and have had two or three years practical experience in using SS.

The medium or smaller sized firm are unlikely to employ a member of staff having this experience. Some form of programme for training SS project co-ordinators is needed but is not available at present. Perhaps a secondment to a larger firm experienced in the use of SS could be arranged?

We will return to this point in our discussion of training in SS.

11.3. <u>Ensure access to a statistical expert</u>.

Although the individual responsible for co-ordinating SS need not have a statistical background it is useful for any firm contemplating the use of SS to have access to a qualified applied statistician if statistical problems should arise.

The role of the statistician in this case is identical to the role of the legal expert or the tax specialist. His advice may rarely be required but it is important that such an expert is known to be available if the need arises for specialist advice.

It is relatively easy for a medium or small firm to find

a legal or tax expert but not so easy to find an applied statistician who is knowledgeable about SS.

The large international firms can usually rely on their own statistical experts in the management services division. Medium and smaller firms are unlikely to enjoy this facility.

We would, therefore, urge that a roster of applied statisticians who have worked on SS should be prepared by the Accounting Institutes and made available to members seeking advice on this matter.

11.4. Devise appropriate documentation.

SS must be systematically applied according to a strict set of rules. This requires the user of SS to make available to the audit staff:

1. A booklet setting out the rules to operate SS in this particular firm.
2. A set of documents to record such things as confidence level, upper bound on error value, sample size, items drawn in sample, errors discovered, conclusions, as well as the conventional information, client's name, population being audited, auditor's name and so on. Several international firms have devised well planned documentation for recording SS results. Some of these are available on request. Naturally the

documentation is determined by the system of SS in use.

The clear documentation of audit method and results which is required by SS is one of the strongest arguments favouring the method.

Other methods may provide as clear an audit record but they need not do so.

11.5. Ensure that a suitable teaching programme is available. All of the large auditing firms interviewed by us provided their own teaching programme. None of these firms used external courses such as those run by the various professional institutions. (See Chapter 8).

The length of the teaching programme depends on the type of SS being used. SS systems using attribute sampling to estimate an upper bound on error rate can be taught in one day of six hours. SS systems which attempt to estimate an upper bound on error value need at least three days (18 hours) but three firms run courses exceeding 50 hours.

There is a need for a properly designed series of courses on SS for the medium sized and smaller firms. These should teach both sampling for error rate and sampling for error value.

We expand on this topic in Chapter 12.

Even when an external course is available to teach the theory and practical application of SS to auditing, an in-house company course will still be needed to teach how SS is to be integrated into the existing current method of the individual firm.

This material can be covered in a one day course of around eight hours.

11.6. Acquire an SS computer package.

Computer packages are now available to implement all of the systems of SS discussed in this book.

A few of these are available from independent software houses but most have been developed by large professional accounting firms.

We understand that most firms are prepared to lease their computer packages for the requisite fee.

It would not be an economic proposition for a medium or smaller size accounting firm to develop its own computer package.

Since the number of viable SS systems are few, the various

Accounting Institutes might consider compiling a library of SS computer packages and leasing them to individual user firms.

11.7. Integrating SS into existing audit procedures.

When an audit firm decides to use SS it will already have a framework of audit procedures. How will SS be integrated into these procedures?

Our study of this problem in both England and the United States suggests that the integration of SS into existing procedures is not a major problem.

The framework of planning, analytical review, sampling, recording and evaluating remains unchanged. The major changes lie in the rigour demanded in the implementation of the audit. Objectives, confidence levels and the level of accuracy required needs to be stated very precisely. The sample size is calculated objectively and precise benchmarks for acceptance or rejection laid down in advance of the audit sampling.

In the early days of SS auditors, accustomed to the more subjective approach of traditional sampling, may resent the additional rigour imposed by SS. They may interpret this rigour as a lack of flexibility.

Once the sample is drawn the actual testing process is no

different to testing a sample drawn by traditional methods. However the conclusions to be drawn from this test have been laid down in advance.

In some ways this takes a burden off the auditors shoulders. The audit conclusions and subsequent action follow automatically from the result of the test. The auditor need not ask "What do I do now?"

The results and conclusion of the audit test will be recorded in a standard format and this allows an easy comparison of the results of different audits.

The standard format of SS audit results makes it rather easier for managers and partners to control and review the audit.

11.8. Summary and conclusions.

If an accounting firm decides to adopt SS it must devise a formal plan for integrating SS into its organisation and procedures.

The more important aspects of this plan will be a formal policy decision at the highest level to use SS for a fixed period before review and the preparation of suitable documentation and training programmes.

There is some controversy as to which method of SS should be adopted initially. We advocate the initial adoption of acceptance sampling to test compliance errors. Later, after, say, three years or so, the MUS system can be introduced to test for value error.

The introduction of SS should be supervised by a specific individual and not allocated to a level of staff, say audit seniors. An expert on SS should be identified in the firm. This individual will handle specific problems in application as they arise.

It is advisable to arrange access to a statistical expert, possibly outside the firm, but experience suggests that expert statistical advice is rarely needed.

There is a need for proper documentation. We suggest that a firm introducing SS should use existing documentation devised by one or other of the larger firms using the given method of SS. This, we are informed, would be made available at minimum or zero cost.

The lack of proper training facilities in SS is a problem. Several large firms indicated to us that they would consider providing access on their courses to the staff of medium sized firms who wished suitable training in SS. Alternatively the various Accounting Institutes may provide adequate courses in the near future.

Computer software on SS is now available on time sharing and micro-computers.

Our experience in introducing SS suggests that integrating SS into existing audit procedures is not a major problem.

Chapter 12.

The training of accountants in statistical sampling.

12.1. Sources of information on SS.

The facilities for training accountants in the use of SS in the United Kingdom are not satisfactory. The large accounting firms provide in-house training of good quality but it is difficult for a medium sized accounting firm to find adequate training facilities on SS in England. (1).

Training material in SS is available from several sources:

(A) Books, Doctoral dissertations and journal articles.

(B) Course manuals.

(C) Video-tape courses.

(D) Lecture courses from (a) accounting institutes (b) other training courses.

(E) In-house training courses by large accounting firms.

12.2 Books and articles.

At least fifteen books and several hundred articles have been written on applying statistical sampling to auditing! We attach an annotated bibliography of some of these books and articles in Chapter 15.

Several of the books provide an excellent introduction to the statistical theory underlying SS and a useful discussion of the basic philosophy of auditing but are of limited value to the auditor who wants to know "how to do it in practice." They are good on theory and concepts but are not very useful as training manuals. They are not designed as such. Perhaps the one book which might be considered an exception to this rule is D. Leslie and others (1980). This book was based on a training manual, which may explain its more pragmatic approach.

Many books on SS seem to us to be stronger on statistics than on auditing. Statistical sampling can only be learned by doing, and reading books on the subject is no substitute for applying SS to a real life audit. In this respect Vanasse (1968) is useful since he provides a set of pseudo-accounting populations to audit. The reader can apply what he has learned of SS to calculate the required sample size, draw a random sample, and estimate the % of error or the value of the population being audited.

Roberts (1978) is strong on describing the mathematics of SS and McRae (1974) discusses some of the practical problems in applying SS. Newman (1976) provides a wide selection of non-audit applications of SS based on computer sampling. With regard to tables, Arkin (1974) provides the most comprehensive set of sampling tables for acceptance sampling, discovery sampling and variables sampling. Leslie and others (1980) provides the table needed for dollar unit sampling. Smith (1976) introduces the statistics used in auditing.

12.3 Doctoral dissertations.

Several doctoral dissertations have been presented on the subject of statistical sampling in auditing. Some of these provide material helpful to practitioners of SS.

Guinn (1973) provides a useful discussion of the legal implications of SS and describes the results of a survey of the use of SS by CPA firms in four states of the USA. He describes the problems encountered and some factors limiting the use of SS in auditing.

Dennis (1972) describes the development of SS as an audit tool. He also provides the results of a survey of the use of SS by auditing firms in the USA on auditing debtors.

Joseph (1972) provides a useful summary of the early investigation of SS by the AICPA. He also provides a survey of the use of SS by those CPA firms in the United States having two or more AICPA members. He provides useful information on the attitudes of auditors to statistical sampling.

Newman (1972) describes the application of SS to estimating the value of inventory. The method requires the use of a computer.

Laudeman (1976) examines the relationship between precision and materiality. He considers that this relationship should not be inflexible, as it is with many SS methods. He suggests a method of providing a flexible relationship. Pushkin (1978) examines the relationship between nominal and true levels of risk in audit sampling. The dissertation provides a useful discussion of alpha and beta risk in relation to auditing. A method is provided for converting nominal alpha risk to true alpha risk.

Vanecek (1978) provides a careful analysis of the Bayesian approach to SS in auditing. He concludes that the method is most efficient when the audit sample size is small and the error rate is low. He finds the DUS method very convervative for beta risk and advocates a sensitivity analysis approach to statistical auditing.

12.4. Publications by professional accounting institutes.

Several of the professional bodies of accountants have published papers relating to SS. Most of these publications only touch on SS indirectly.

We provide a list of these publications in Chapter 15. Several publications are now rather dated. By far the most important are the appendices to the "Codification of statements on auditing standards" published by the AICPA.

12.4.1. Course manuals.

The education division of the American Institute of Certified Public Accountants provide a wide range of training manuals on SS. These can be divided into two categories, self-tuition courses and course manuals for running lecture courses on SS. The latter consist of an instructors manual with detailed teaching instructions plus manuals for course participants.

The self-tuition manuals are in the form of programmed learning texts. These are available on statistical concepts related to SS, sampling for attributes, stratified sampling, ratio estimates and a useful Field Manual for SS. These publications are listed in Chapter 15. They are available from the AICPA, Continuing Education Programme, New York.

The AICPA course manual on SS consists of a sixteen hour programme of teaching materials, problems and case studies, entitled "How to use SS sampling for attributes on smaller companies" by A.A. Arens. The instructors manual is very thorough.

This course manual is well designed for the medium sized company who wish to use SS for compliance testing.

12.4.2. Video tape courses.

An excellent video tape course on monetary unit sampling has been designed by the Canadian accounting firm Clarkson Gordon of Toronto. The course can run for eight hours and provides information on both the statistical theory behind SS and examples of practical application. A course manual is also available. This course has been well tested and is subject to continual revision to incorporate new developments in theory and practice.

The AICPA continuing education programme have also developed a number of short cassettes (non-video) programmes of self-tuition on SS. These courses cover the basics of sampling theory as applied to attribute and variables sampling.

12.5 Lecture courses.

(a) Run by the professional institutes.

Lecture courses on SS have been run by the professional institutes in the United States since the late fifties. In the UK and Canada since the mid-sixties and in South Africa since 1970. There has been no shortage of courses but we would question whether these courses provide sufficient information to allow a firm to introduce SS into its audit procedures.

We have examined the content of several of the courses run in the UK and most of them adopt an introductory or appreciation approach. They appear to be designed to introduce the concept of sampling theory and emphasise the advantages of SS but they do not provide sufficient training in the practical operation of the techniques.

The only course we found which devoted sufficient time to the practical operation of SS was the course run by Arthur Young for the ICAS in November 1979. This three day course provided sufficient information on the monetary unit sampling method to allow a participant to go out and use it in practice.

There is urgent need of several different levels of courses on SS in the UK. We suggest mounting the following courses:

(1) An appreciation course on the advantages of SS. (one day).

(2) A technical course on the basic statistical theory behind SS. (one day).

(3) A course on attribute sampling for compliance testing. (one day).

(4) A course on monetary unit sampling. (three day).

(5) A seminar on practical problems of applying SS and recent developments in SS. (one day).

Courses could also be run on stratified variables sampling and computer assisted SS if there was sufficient demand. The courses would be complementary, for example, participants would be expected to attend course (2) before attending course (4).

The Institute of Internal auditors have run a few one day appreciation courses on SS and the City University Business School include SS in their internal audit programme.

We conclude that inadequate provision is made in the UK, on publicly available courses for training auditors in the use of SS.

(b) In house courses.

The situation with regard to in-house courses run by the large professional firms is quite different. We examined several of these courses and most were of good quality. They laid great stress on practical application and most included case studies. Their main weakness was a tendency to teach a mechanical approach to SS without explaining the underlying statistical theory.

Unfortunately only one of those in-house courses is publically available (2) although many firms make the course available to their clients internal auditors.

Perhaps the larger accounting firms could be encouraged to open up their courses to staff of medium sized firms.

We include a list of courses in Chapter 15.

12.6. Conclusions.

There is no problem in finding books and articles on SS. Several excellent textbooks are available and there is a plethora of journal articles on the theory of SS. Perhaps more articles are needed on the practical problems of applying SS.

At least one good course manual and one excellent video-tape course on SS are available at reasonable cost. Lecture courses on SS are inadequate at present. The in-house courses run by the large professional accounting firms are good but not available to the general public. The courses run by the professional institutes lack depth, particularly on the operational side.

We suggest that a series of courses on SS at various levels should be sponsored by the professional institutes.

Notes on Chapter 12.

1. The Arthur Young course described later was run by the ICAS in Scotland.
2. The Clarkson Gordon course which is, we believe, the basis for the Arthur Young course run in Scotland in November 1979.

Chapter 13

The computer and statistical sampling.

13.1 Introduction

Auditors often ask the question "Is a computer needed to operate SS?".

The answer is no. A computer is not necessary for attribute sampling (and this includes the MUS method). A computer is needed to operate variables sampling with stratification.

Although a computer is not a necessary condition of attribute (error rate) sampling, it is very useful if available. If a computer is available, and if the accounting population being audited is stored in a computer, the use of SS is much facilitated.

If the accounting population to be audited is stored on a computer there is a natural bias towards using scientific sampling.

The computer will almost certainly be programmed to list a set of items to be audited, so why not use a random sampling technique rather than any other method of selection? Also traditional judgement sampling is rather more difficult to employ when the accounting population to be audited is stored in a computer. Finally the calculation of sample size and any other statistical calculations are performed rapidly and accurately by the computer.

There is, then, a natural bias towards SS when the data are stored on a computer.

The computer can be programmed to carry out the following operations:

(1) Generate a given set of random numbers.

(2) Given (a) confidence level (b) required precision (c) expected error rate or standard deviation, the computer can calculate the required sample size.

(3) Given the sample size calculated under (2) the computer can use the random numbers generated under (1) or some other rule to draw a random sample of required size from the given population.

(4) Once the auditor has audited this sample the results of the audit can be fed back to the computer for analysis. The computer can examine the sample results and provide an opinion on the next course of action, i.e. accept population, extend the audit, try another sampling method or reject the population.

It should be emphasised that most SS methods are so simple to operate that the computer need only be used for selecting the list of items to be audited.

13.2. Whose computer?

An important distinction needs to be made between the situation where an auditor uses his own computer to assist with SS and those cases where he uses the client's computer.

Activities (1) (2) and (4) above can be carried out on the auditor's own computer but the drawing and printing of the audit sample will normally be effected by using the clients computer. This activity will require some coordination with the client. The sampling process can often be performed in tandem with the periodic run of the client's tape or discs on a normal month end run. Thus the incremental cost to the client can be negligable.

An alternative possibility is for the auditor to run his client's tapes on his own computer, but this requires the two systems to be compatible, or that they can be made compatible.

Further discussion of the mechanics of sampling by computer would carry us into the various philosophies of computer audit. We will not pursue the matter further here.

13.3. Future developments.

It may be that the power of the computer will become an indispensable aid to the future development of SS.

For example the calculation of a less conservative upper bound on the MUS method may need to be delegated to the computer on each audit. Also, if it is decided that no one method of SS is suited to all audits then the computer may be programmed to select that sampling method best suited to this particular audit.

Further research in auditing may reveal specific error patterns in specific types of accounting populations (see Ramage and others (1979)). If this is so the computer might be programmed to analyse errors to identify the existence of a given pattern and then select that sampling plan best suited to detecting this error pattern while being most economical on sample size.

It is possible that developments lie ahead in this area of auditing, but much more research is needed before these ideas become operational.

There can be little doubt that the introduction of the micro-computer, costing a few thousand pounds, has brought computing power within the grasp of all but the smallest firms. Within a few years virtually all audited accounting populations will be stored in a computer. This must surely lead to the universal use of computer audit packages and scientific methods of audit sampling? (1)

13.4. Current state of play

Every one of the ten large firms using some variant of SS had developed a computer audit package which incorporated a statistical sampling sub-routine.

These packaged programmes performed the set of activities set out above. Our somewhat superficial study of these programmes suggested that they were, if anything, more powerful than they needed to be. We can put this another way by saying that the full power of these SS programmes was seldom used.

The reason for this state of affairs is easy to explain. The auditors using the packages were seldom fully aware of the power of the programme available to them. The training programmes on SS by computer appear to concentrate on the mainstream uses of the programme and provide a somewhat limited introduction to the more esoteric aids so assiduously built in by the programmers.

In most of the cases studied by us the auditors were well trained in the mainstream use of the computer for SS work and the integration of the computer into the audit routine was smoothly handled.

There appears to be no major problem in using the computer as an audit tool once the initial set up problems (such as modifying clients files) are overcome.

Accounting firm	Name of package or programme
Arthur Andersen	AUDEX
Arthur Young Mc.lelland Moores	AYSHARP/AYSTRATIFY
Deliotte, Haskins Sells	AUDITAPE II CARS III
Peat, Marwick Mitchell	SYSTEM 2190
Price Waterhouse	PWAAA
Thomson McLintock	RATEST
Touche Ross	TRSS

Exhibit 13.1

Some audit computer packages incorporating SS techniques.

Computer packages incorporating statistical sampling techniques have been developed by many of the larger accounting firms. The names of some of these packages are listed above.

Roberts (1978) provides a detailed guide to a series of computer packages designed to assist in applying SS to audit populations. These packages written in COBOL and BASIC are made available for a nominal charge from the AICPA (Computer Services Division).

All of the above comments apply to the <u>major</u> audit firms in the U.K.

Among the medium and smaller firms we located few who were using SS. Only two of these used a computer. The explanations for not using a computer were the obvious ones of expense, expertise, and accessibility. The large firms can use a computer because they almost all have departments specialising in this field. The medium and smaller firms do not enjoy access to such a facility.

We feel that the spread of micro-computers will alter this situation but the smaller firms will need help. The Accounting Institutes can, perhaps, provide some kind of central service facility connected by telephone to the smaller audit firms? We will return to this point in our conclusions and recommendations.

3.5 Summary and conclusions.

The main impetus behind the adoption of SS in auditing is the increasing use of the computer in accounting work.

If an audit sampling procedure must be programmed, why not use a scientific sampling procedure?

SS can be, and is, widely used without the help of a computer, but if a computer is available it can provide considerable assistance to the auditor.

Many computer audit packages now incorporate an SS subroutine.

The computer can be programmed to calculate audit sample size, draw a random sample and evaluate the results of the sample. It may even be programmed to suggest the best option for further action given the sample results.

Exciting possibilities lie ahead in combining the statistical analysis of sample data drawn from a computer store with an auditor's judgement and experience. But we need to know more about standard error patterns in accounting data for these possibilities to be applied in practice.

The increasing use of microcomputers is likely to have a major impact on the use of SS in the decade. (1)

Notes.

1. For example the STAR system of analytical review costs at least \$50 on a time shared large computer but only \$15 per run on a microcomputer. D.H.S. "Auditing and accounting newsletter". 22nd December 1980. p.7. An experienced user of STAR claims those figures to be too high. His figures were in the region of \$2 to \$3 per run.

Chapter 14

Legal aspects of applying statistical sampling to external auditing.

14.1. Can statistical audit samples be used as legal evidence?

If an auditor selects an audit sample using statistical methods of selection can this sample be used as evidence in a court of law and, more particularly, would this statistical sample carry greater evidential weight than a traditional sample?

This is a good question which demands an answer. Unfortunately the legal literature on statutory and case law fails to provide us with an answer. An audit sample would almost certainly be admissable as evidence (1) but the adequacy of the audit sample has not been tested by the English legal system. There is no case law on the subject. Statutory law does not attempt to define the meaning of an adequate audit test let alone an adequate audit sample size.

Those few legal cases which have examined the adequacy of audit performance by external auditors have tended to look to other auditors to measure the adequacy of audit performance. No objective measure of audit performance has been attempted. The techniques of sample selection and the adequacy of audit sample size have not been argued before the English Courts.

This unsatisfactory state of affairs exists not only in England and Wales but also it seems in Scotland, the United States, Canada and the Netherlands. We could find no legal case anywhere which had attempted to define an adequate audit sample. (2).

The legal obligation of external auditors in England arises from statutory obligations under the various Companies Acts and common law.

Neither the companies acts nor common law provides guidance as to the techniques of audit sampling to be used or the adequacy of sample size. The auditor is simply required to audit in accordance with generally accepted audit standards. These standards are qualitative not quantitative.

Several audit firms interviewed by us stated that it was hoped that SS might provide an objective quantitative standard which should prove useful as evidence of competancy in a court of law if the firm should be accused of negligence. As yet this expectation has no support in law.

As Guinn (1973) has pointed out it is difficult to interpret qualitative standards through quantitative procedures but "several CPA firms .... have become interested in using the techniques (of SS) as a means of obtaining sample data that is defensible in court". (page 56)

Sir Richard Eggleston (1978) in his book on the legal implications of statistical theory suggests that where legal evidence is concerned "there is no reason why sampling techniques should not be followed to save expense .... if the probability level obtained by the sampling method is reasonably high". (page 184). He continues, "It will be apparent from what has been said in this book that where there are available statistics on which an expert witness can forward an opinion, there is nothing to prevent him from applying acceptable methods of calculating probabilities and presenting the results to the court. But it will also be apparent that on the whole the courts have not been receptive to attempts to use experts in any way which might deprive the judges or juries of their traditional function in relation to fact finding or prediction". (page 186)

..... and here the matter would appear to rest so far as the English courts are concerned.

14.2. The situation in the United States.

The evidential value of statistical samples in auditing has not been tested in a court of law in the United States. However scientific sampling using accepted statistical methodology has been discussed in U.S. legal circles for over twenty five years. See Sprowls (1957).

The U.S. courts have decided that samples can be used as evidence to provide information about populations from which they are drawn (ESCOTT v BARCHRIS 62 CIV 3539 March 1968). However the auditor has the burden of proving that the extent of his examination is sufficient.

When samples are presented the court has emphasised the importance of proving homogeneity in the populations from which they are drawn. For example in Sears Roebuck v City of Inglewood (Los Angeles 1955) statistical sampling was rejected in a non-audit case possibly because of the lack of homogeneity in the population (overpayment of sales taxes). See Sprowls 1957.

Statistical sampling has been frequently used in cases brought to court under the Federal Food and Drugs Act. The populations here are thought to be relatively homogeneous.

A further factor limiting the use of statistical samples as legal evidence is that the courts require proof that the population from which the sample was drawn was 'inclusive'. Statistical samples have been rejected in several opinion surveys because the courts doubted whether the population sampled was inclusive of all relevant persons.

Inclusion is not likely to prove a major problem to the auditor but homogeneity is a major problem in applying SS to accounting populations.

There can be little doubt that audit papers are admissable to the courts. They avoid exclusion as hearsay evidence since they are both business records and records to support recollection of past events (like a policeman's notebook). Note that external audit papers, unlike say private medical records, can be called before the court in the U.S.A.

It seems that if sample data is presented as evidence before the court this sample data carries greater weight as evidence if the sample is selected and evaluated using accepted statistical method. In the case of

State Wholesale Grocers v Great Atlantic and Pacific Tea Company (154 F Supp. 471 (1957)) the judge admitted the statistical sample on the basis that the sample was selected in accordance with scientific sampling method.

The American Bar Association has encouraged the use of statistical sampling, and classes on scientific sampling methods are currently being held for lawyers in New York. (3)

The admissibility of sample data into a court of law in the United States is discussed by McCoid (1957), Kecker (1955), Sprowls (1957), Wigmore (1940), Dean (1954) and Guinn (1973).

W. Edwards Deming, the noted applied statistician, has written (Deming 1954) that:

"Wider use of standards of statistical performances and standards designed especially for surveys in accounting and in legal evidence, would eliminate some of this needless waste and delay, and they would also help to eliminate some of the wrong judicial decisions that are made (a) on the basis of bad statistical evidence, and (b) on failure to recognize good statistical evidence."

A statement of standards of sampling for legal evidence was published by the U.S. Society of Business Advisory Professions in 1954.

Copeland and Englebrecht (1975) provide a careful discussion of the weight of statistical audit samples as legal evidence. They conclude that such samples may improve the defence against negligence but that the legal arguments are difficult and fragile.

C. and E consider that, "Auditors are in a stronger defensive position with a sample that has been scientifically drawn. (p.24). They conclude that, "audit sample size is one item upon which statistical methodology can provide conclusive evidence."

C. and E make the interesting suggestion, not mentioned elsewhere in the literature, that the auditor should carry out a sensitivity analysis on the parameters of the sample size calculation to test the sensitivity of the size of sample to changes in the factors which determine sample size. (i.e. standard deviation). Otherwise C and E consider that the auditor may be open to attack <u>on statistical grounds</u> in the courts.

Randall and Frishkoff (1976) also examine the legal implications of SS applied to external auditing.

They provide a synopsis of most of the recent legal cases in the USA which consider the application of statistical method to evidential matters.

They claim that the US inland revenue service "explicitly condones only <u>probability</u> sampling" in certain of its standard procedures. (p.99). See note 4.

"A bill approved in 1964 by the (US) House of Representatives specifically permitted the use of SS procedures in the examination of vouchers." (p.100.)

Ijiri and Leitch (1980) suggest using information found in populations similar to the one under audit to improve the auditors estimate of the error rate. This approach would seem to improve the auditor's legal defence against negligence by providing an empirical foundation for his judgement as to the "reasonableness" of the error rate or value found in the sample.

Perhaps it is appropriate to finish this Chapter on a cautionary note by quoting the conclusions of Randall and Frishkoff (1976). They conclude with the statement that; "The courts are relying more heavily on sophisticated statistical documents whose standards barely condone... non-random sampling when probability sampling is feasible." "We caution those auditors who continue to use judgement sampling in the presence of feasible probability sampling procedures to be prepared to defend their logic in so doing." (p.100).

14.3 Conclusions

We conclude that the evidential weight of scientific samples in auditing is not proven. The current lack of knowledge of statistical theory in the legal profession is likely to prove a severe barrier to drawing evidential weight from such samples.

There appears to be no barrier to presenting statistically derived samples as evidence, but ouside the United States we doubt if at present (1981) such samples would carry greater evidential weight than traditional audit samples. Even in the United States, where the courts appear to give more weight to statistically derived samples, we could find no evidence that conclusions derived from statistically based audit samples would be given more evidential weight than conclusions derived from traditional audit samples.

However, we caution the auditor who believes that the current legal situation is permanent. It appears to us from an examination of current legal journals that statistical thinking is beginning to permeate the legal profession and that, at some point, in the not too distant future the law may require samples to be drawn scientifically if they are to be presented as evidence in the courts.

Notes on Chapter 14.

1. See Eggleston (1978) Chapters 6 and 14.

2. The adequacy of sample size in auditing was touched on in SEC release number 15978 June 27th 1979 and Accounting Series Release No. 153A June 1979 in the United States. The case concerned the audit of the Giant Stores Corporation and Ampex Corporation.

"The commission finds the audit programme for the advertising credits was inadequate because of the limited sample selected for external verification."

"The audit sample selected was inappropriately small and the responses received were too few to provide adequate audit evidence."

(The Auditor) "should have selected a much larger sample than 50 of several thousand vendors to check the accuracy of Giant's accounts payable."

A statistical approach would have clarified the adequacy of the sample size.

3. Information supplied by New York Bar Association.

4. See Revenue Procedure 64 - 4 of the US Internal Revenue Service and the important appendix compiled by two noted statisticians (DEMING and GLASER).

Chapter 15.

A review of the literature.

Introduction

Published material on the application of SS to audit work is extensive. We traced fifteen books and over five hundred acticles in the professional and academic journals.

The following chapter will attempt to provide the reader with a guide to this extensive literature.

The chapter is divided into the following parts.

Coding.

1. Books
2. Doctoral dissertations.
3. Sets of articles on selected topics.
4. Publications by the Professional Institutes.
5. Course manuals.
6. Video tape course.
7. Lecture courses.
8. Selective bibliography.

The bibliography lists all of those articles cited in the study.

In our opinion this bibliography includes virtually every important article written on the subject of applying SS to external auditing.

1. Books on SS applied to auditing.

1.1 ARKIN H.

"Handbook of sampling for auditing and accounting" (2nd Ed) New York, McGraw Hill 1974.

(A clear introduction to sampling theory plus an excellent set of sampling tables)

1.2. COEN C.D.

"Statistical sampling for bank auditors"

Park Ridge. Illinois. Bank Administration Institute 1973.

(provides examples of SS applications in banking)

1.3. CYERT R.M. and DAVIDSON H.J.

"Statistical sampling for accounting information"

Englewood Cliffs. New Jersey, Prentice Hall 1962.

(An early general introduction to SS with lots of useful examples and solutions)

1.4. LESLIE D., TEITLEBAUM, A and ANDERSON D.

"Dollar unit sampling"

Pitman, London 1980.

(Possibly the best book on SS yet written. Concentrates on the monetary unit sampling approach but discusses other methods. Provides the key sampling table for MUS and evaluation procedures for examining sample results)

1.5. McRAE T.W.

"Statistical sampling for audit and control"

(Describes briefly all the techniques including MUS. Rather more on problems of practical application than other texts).

1.6. MEIKLE G.R.

"Statistical sampling in an audit context"

Canadian Institute of Chartered Accountants.

Toronto 1972.

(Explains SS with emphasis on MUS type approach. Useful in placing SS in a broader context and discussing the pitfalls to avoid).

1.7. NETER J. and LOEBBECKE J.K.

"Behaviour of major statistical estimators in sampling accounting populations - an empirical study."

AICPA - New York 1975.

(A test of various SS systems assuming different kinds of error rates.)

1.8. NEWMAN M.S.

"Financial accounting estimates through statistical sampling by computer" John Wiley - New York 1976.

(Describes a wide range of uses of SS including application to auditing)

1.9. ROBERTS D.M.

"Statistical auditing"

AICPA - New York 1978.

(Possibly the best academic book on SS. Provides a careful but perhaps too concise description of each method of SS. A case study is included and description of suitable computer packages. Includes admirable glossary, bibliography and list of mathematical formula for each method (but without describing meaning of symbols!)

1.10. SMITH T.M.F.

"Statistical sampling for accountants". Haymarket. London. 1976.

(An excellent introduction to the statistical theory behind basic audit sampling).

1.11. VANASSE R.W.

"Statistical sampling for auditing and accounting decisions - a simulation"

McGraw Hill. New York 1968.

(A set of pseudo-accounting tables useful for practice in using SS techniques. Since they are not totalled MUS cannot be used. Very useful book for teaching basic sampling principles. Includes sampling tables and basic description of acceptance, discovery and variables sampling).

1.12. Vance L.L. and Neter J.

"SS for auditors and accountants" John Wiley. 1956.

2. <u>Doctoral dissertations</u>.

Several doctoral dissertations, all from US Universities, have studied various aspects of applying statistical methods to audit sampling. Some of the more useful of these are listed below.

2.1. DENNIS D.M.

An investigation into the use of SS.

University of Missouri - Columbia 1972.

2.2. GUINN R.E.

Statistical sampling : a study and evaluation.

University of Alabama 1973.

2.3. JOSEPH J.

Sampling and the independent auditor.

University of Oklahoma 1972.

2.5. LINDBECK R.S.

"An enquiry into the suitability of non-parametric statistical tests for accounting and auditing problems."

University of Alabama 1969.

2.6. NEWMAN M.S.

Statistical estimate of computer based inventories.

New York University 1971.

2.7. PUSHKIN A.B.

An investigation of the validity of auditing procedures used in mean-per-unit sampling plans.

Virginia Polytechnic Institute and State University. 1978.

2.8. VANECEK M.T.

Bayesian Dollar Unit sampling in auditing.

University of Texas at Austin. 1978.

All of these dissertations are available through University Microfilms International, London, U.K.

3. Sets of articles on selected topics.

The articles listed on the following pages are well written introductions to specific topics on SS for those readers who wish to pursue these topics in more depth.

3.1. Objectives

SS must always be placed in proper perspective by relating the method to the objectives of external auditing. This subject is discussed fully in the following articles.

1.1 Elliott, R.K. and Rogers, J.R., "Relating Statistical Sampling to Audit Objectives", The Journal of Accountancy, July 1972, pp. 46-55.

1.2 Ijiri, Yuji and Kaplan, R.S., "A Model for Integrating Sampling Objectives in Auditing", Journal of Accounting Research, Spring 1971, pp. 73-87.

3.2. The theory of SS

The theory of attribute sampling for compliance testing can be found in any elementary textbook on statistical theory. The theory behind substantive testing of value is rather more difficult. The following articles are mainly concerned with calculating an upper bound on error value.

3.2.1 Boer, Germain, "Role of Judgment in Statistical Sampling", CPA Journal v. 44, March 1974, pp. 39-43.

3.2.2 Garstka, S., "Models for Computing Error Limits in Dollar-Unit Sampling", July 1976. Symposium on Frontiers of Auditing Research, University of Texas at Austin, April 1976.

3.2.3 Feinberg, J., Neter, J., Leitch, R., "Estimating the total overstatement in accounting populations", Journal of the American Statistical Association, June 1977.

3.2.4 Kaplan, R.S., "Statistical Sampling in Auditing with Auxiliary Information Estimators", Journal of Accounting Research, Fall 1973, pp. 238-258.

3.2.5 Loebbecke, J.K. and Neter, John, "Considerations in Choosing Statistical Sampling Procedures in Auditing", Journal of Accounting Research, v. 13, Supplement, 1975.

3.2.6 Neter, J., Loebbecke, J.K., "On the behaviour of statistical estimators when sampling accounting populations", Journal of the American Statistical Association 1972, pp. 501-507.

3.2.7 Roberts, Donald M., "Statistical Interpretation of SAP No. 54", Journal of Accountancy, v. 137, March 1974, pp. 47-53. (Examines the problem of balancing type 1 and type 2 errors in audit sampling)

3.2.8 Teitlebaum, A.D., "Dollar-unit Sampling in Auditing", (paper presented to National Meeting of the American Statistical Association, Dec. 1973)

3.3. Standards

It has been argued that statistical standards can and, perhaps, ought to be set for external auditing.

3.3.1 Broderick, John C., "Setting Standards for Statistical Sampling in Auditing" (in Arthur Andersen/University of Kansas Symposium on Auditing Problems, 1974, Contemporary Auditing Problems, Lawrence, Kan. 1974, pp. 77-84).

3.3.2 Stringer, Kenneth W., "Toward Standards for Statistical Sampling", (in Touche Ross/University of Kansas Symposium on Auditing Problems, 1972. Auditing Looks Ahead, Lawrence, Kan., 1972, pp. 43-9).

3.4. Methods of SS available

Various statistical methods have been applied to audit sampling. the following articles describe the various methods.

Regression analysis

3.4.1 Deakin, E.B., Granof, M.H., "Regression analysis as a means of determining audit sample size", Accounting Review, Vol. XLIX No. 4, October 1974, p. 764.

3.4.2 Gurry, E.J. and Santi, D.W., "Regression analysis as an audit technique", The Australian Accountant, April 1975, pp. 132-139.

3.4.3 Newman, M.S., "Regression estimates for Accounting Purposes", (in Haskins & Sells Selected Papers, 1972. New York, 1973, pp. 172-94).

Ratio and difference estimates

3.4.4 Akresh, Abraham D., "Use of the Ratio Estimate in Statistical sampling - A Case Study" (Accounting and Auditing), New York Certified Public Accountant, v. 41, March 1971, pp. 221-4.

3.4.5 McCray, J.H., "Ratio and Difference Estimation in Auditing", Management Accounting (NAA), v. 55, Dec. 1973, pp. 45-8.

Monetary unit sampling

3.4.6 Anderson, R.J. and Teitlebaum, A.D., "Dollar-Unit Sampling", CA Mazagine, April 1973, pp. 30-38.

3.4.7 Goodfellow, J.L., Loebbecke, J.K. and Neter, J., "Some Perspectives on CAV Sampling Plans", CA Magazine, Oct. 1974, pp. 22-30 and Nov. 1974, pp. 46-53.

3.4.8 Teitlebaum, A.D., Leslie, D.A. and Anderson, R.J., "An Analysis of Recent Commentary on Dollar-Unit Sampling in Auditing", available from authors on request. (Clarkson Gordon, Toronto, Canada).

Acceptance sampling

3.4.9 Finley, D.R., "Controlling compliance testing with acceptance sampling", CPA Journal, vol. XVIII No. 2, Dec. 1978, pp. 30-35.

Decision theory

3.4.10 Kinney, Wm. R., "Decision-theory Approach to the Sampling Problem in Auditing", Journal of Accounting Research, v.13, Spring 1975, pp. 117-32.

3.5. Bayesian theory and auditing

3.5.1 Corless, J.C., "Assessing prior distributions for applying Bayesian statistics in auditing", The Accounting Review, July 1972, pp. 556-566. (See also A.R. January 1975 for a subsequent discussion)

3.5.2 Crabtree, M.G., "New approach to effective sampling for auditors", Accountant 29 April 1976, p. 500.

3.5.3 Knoblett, James A., "Applicability of Bayesian Statistics in Auditing", Decision Sciences, v. 1, July-Oct. 1970, pp. 423-40.

3.5.4 Kraft, W.H., "Statistical Sampling for Auditors: A New Look", The Journal of Accountancy, August 1968, pp. 49-56.

3.5.5 Tracy, J.A., "Bayesian Statistical Methods in Auditing", The Accounting Review, January, 1969, pp. 90-98.

3.6. SS and the small firm

6.1 Mottershead, A., "The small company audit - the answer?", Accountancy February 1978.

3.6.2. Naus, J.H., "Effective uses of SS in the audit of a small company." The Practical Accountant (US) March/April 1978. p.33.

3.7. Errors in accounting populations.

3.7.1. Taylor, Robert G., "Error Analysis in Audit Tests" (Accounting & Auditing). Journal of Accountancy, v. 137.

3.7.2. Ramage, J.G. and others. "An empirical study of error characteristics in audit populations" Journal of Accounting Research. Vol. 17. (Sup) 1979. pp. 72-113. (A useful study of actual error rates found in external audit work).

3.8. Practical application of SS

The one area of SS on which not enough has been written concerns the practical problems of applying SS to auditing. A few articles are available.

3.8.1. Boatsman, J.R., Crooch, G.M. "An example of controlling the risk of a Type II error for substantive tests in auditing", The Accounting Review, July 1975, pp. 610-615.

3.8.2. Loebbecke, J.K. and Neter, J., "SS in confirming receivables", The Journal of Accountancy, June 1978.

3.8.3. McCray, J.H. "Many parameters of an audit sampling plan", Ohio CPA Vol. 32, Summer 1973, pp. 75-78½

3.8.4. McRae, T.W. "Integrating Statistical Sampling into Conventional Auditing Procedures", Australian Accountant, v. 7, June 1971, pp. 202-7.

3.8.5. Tummins, Marvin, "Block or Random Sampling? by Marvin Tummins and Earl F. Davis", Internal Auditor, v. 27, May-June 1970. pp. 57-64

3.9. Legal aspects of SS

3.9.1. Copeland R.M. and Englebrecht T.D. "SS, an uncertain defence against legal liability." CPA Journal, November pp. 23-27. 1975.

3.9.2. Randall B and Frishkoff P. "An examination of the status of probability sampling in the courts". Auditing Symposium 3. University of Kansas. (Editor H. Stettler) pp.93-106. 1976.

(Both these US studies examine recent legal cases in US courts which have examined evidence involving statistical theory, particularly sampling theory).

4. Publications relating to SS by Institutes of Professional Accountants.

Discussive publications.

	Institute		Publication
4.1.	AICPA	4.1.1.	Codification of statements on auditing standards. Appendix A. "Relationship of SS to generally accepted auditing standards" and Appendix B "Precision and reliability for SS in auditing" AICPA New York 1977 (This supercedes previous publications by the AICPA on SS)
4.2.	CICA	4.2.1.	"Internal control and procedural audit tests" CICA Toronto 1980
		4.2.2.	"Extent of audit testing" CICA Toronto 1980.

Institute		Publication
4.3. CIPFA	4.3.1.	"The use of SS method in audit work" District auditors society. IMTA 1970.
4.4. ICAEW	4.4.1.	"Auditing standards and guidelines." ICAEW. 1980. Section 203. Audit evidence. Section 204. Internal controls. (Discusses sample evidence but not statistical sampling.) For discussion.
4.5. ICAS		Follows CCAB recommendations as above.
4.6. NIVRA	4.6.1.	Netherlands Institute of Registered Accountants "Statistical sampling in auditing". NIVRA, Amsterdam, 1977.
4.7. NZSA	4.7.1.	New Zealand Society of Accountants. "Control and the nature and extent of audit tests" NZSA - Auckland 1977.
4.8. IFA	4.8.1.	Study and evaluation of the accounting system and internal control." International Federation of Accountants. Exposure Draft No. 6. Audit procedures. P. 20 to 24. (Nothing specific on statistical sampling.)

5. Course Manuals

5.1. ARENS A.A.

"How to use statistical sampling for attributes on smaller companies"

AICPA, New York, Continuing Education Programme.

Programmed teaching course.

5.2. "Auditors approach to statistical sampling"

Vol. 1 Statistical concepts

2 Sampling for attributes

3 Sampling for variables

4 Discovery sampling

5 Ratio and difference estimates

6 Field manual

AICPA, New York, Continuing Education Programme.

6. Video tape courses

6.1. "Dollar unit sampling". An admirable eight hour video tape course has been developed by the accounting firm of Clarkson Gordon, Toronto, Canada. This is not for sale but use of the course can be negotiated.

6.2. "SS for attributes"

"SS for variables"

Cassette and short manual available from AICPA, New York, Continuing Education Programme.

7. Lecture courses.

7.1. ICAS

A three day course on monetary unit sampling has

been held. The first course was run by staff of Arthur Young and was based on the course developed by Clarkson Gordon of Toronto, Canada. ICAS have also run several one - and two day courses on attribute and variable sampling, starting in 1966.

7.2. ICAEW.

Some courses on SS were run in the early 1970's but these have been discontinued.

7.3. Institute of Internal Auditors.

Lecture sessions on SS of around three hours are included in the IIA training courses. One day courses on SS have been run, but not on a regular basis. The content of these courses are of the appreciation type, little information is provided on operational problems.

7.4. City University Business School.

The post experience department provides courses for internal auditors which includes three hour sessions on SS. A course for bank inspectors also includes a session on SS in bank auditing.

8. Bibliography

ABRAMS, J. (1947) "Sampling theory applied to the test audit." New York CPA. Vol. 17. No. 10. pp. 645-652.

AICPA (1964) "Relationship of SS to generally accepted auditing standards." Journal of Accountancy. July. pp. 56-58.

AICPA (1977) Codification of Statements on auditing standards: New York. Appendix A. SS and audit standards. Appendix B. Precision and reliability.

ALDERSLEY, S. and TEITLEBAUM, A.D. (1979). "Rigerous DUS cell evaluation". American Statistical Association. Annual meeting. Washington D.C.

AKRESH, A.D. (1971). "Use of the ratio estimate in statistical sampling". New York CPA. March. Vol. 41. No. 3. pp. 221-224.

AKRESH, A.D. (1979) "The use of statistical sampling in the accounting profession." Report to AICPA.

ALY, H. and DUBOFF J.I. (1971). "Statistical versus judgement sampling". Accounting Review. Vol. 46. pp. 119-128.

ANDERSON, R.J. (1977) "The inter-relationship of compliance and substantive verification in auditing." Frontiers of auditing research. University of Texas at Austen.

ANDERSON, R.J. and TEITLEBAUM, A.D. (1973). "Dollar unit sampling" C.A. Magazine (Canada) April. pp. 30-38.

ANDERSON, T.W. and BURSTEIN, H. (1967). "Approximating the upper binomial confidence limit". Journal of the American Statistical Association. September. p. 859.

ARKIN, H. (1961). "Discovery sampling in auditing." Journal of Accountancy. Vol. III. February. pp. 51-54.

ARKIN, H. (1963) (1974). "Handbook of sampling for auditing and accounting" New York. McGraw Hill Book Co.

ARKIN, H. (1976) "Statistical sampling and internal control". CPA Journal. January. pp. 15-18.

BOATSMAN, J.R. CROOCH G.M. (1975). "An example of controlling the risk of a type II error for substantive tests in auditing." Accounting Review, July pp. 610-615.

BROWN, R.G. and VANCE, L.L. (1961) Sampling tables for estimating error rates or other proportions. IBER. University of California, Berkerley.

BURSTEIN, H. (1967). The ratio estimate - a useful sampling technique". New York CPA. November. Vol. 37. No.2. pp. 844-850.

CARMAN, L.A. (1933) "The efficacy of tests" The American Accountant. v.18. December. pp.360-366.

CHARNES, A. and others. (1964). "On a mixed sequential estimating procedure with application to audit tests in accounting." Accounting Review. Vol. 39. No.2. April. pp. 241-250.

CHURCHMAN, C.W. (1952). "Can scientific sampling techniques be used in railroad accounting?" Railway Age. June 9. p.61-64.

CICA (1980) "Extent of audit testing." CICA Toronto.

COCHRANE, W.G. (1963). "Sampling Techniques". John Wiley. New York.

COMPANIES ACT (UK) (1948) and (1967). H.M.S.O.

COPELAND, R.M. and ENGLEBRECHT, T.D. (1975). "SS, an uncertain defence against legal liability." CPA Journal, November. pp. 23-27.

COX, D.R. and SNELL, E.J. (1979). "On sampling and the estimation of rare errors". Biometrika. Vol. 66. No.1.

CRABTREE, M.G. (1976). "A new approach to effective sampling for auditors." Accountant, 29th April. p.500.

CYERT, R. DAVIDSON, H. (1962). "SS for accounting information" Prentice Hall.

CYERT, R.M. (1957). "Test checking and the Poisson Distribution". Accounting Review. Vol. 32. pp. 395-397.

DEAKIN, E.B. and GRANOF, M.H. (1974). "Regression analysis as a means of determining audit sample size." Accounting Review. Vol. 49. No.4. October. pp. 764-771.

DEAN, J. (1954). "Sampling to produce evidence on which courts will rely." Current Business Studies. Vol. 19. Oct.

DEMING, W.E. (1954). "On the contributions of standards of sampling to legal evidence and accounting". Current Business Studies. Vol. 19. Oct. p.25.

DEMING, W.E. (1957). "Standards of probability sampling for legal evidence". Current Business Studies. Vol. 26. p. 3-10.

DEMING, W.E. (1979). "On a problem in standards of auditing from the viewpoint of statistical practice." Journal of Accounting, Auditing and Finance. Vol. 2. No.3. pp. 197-208.

DENNIS, D.M. (1972). "An investigation into the use of SS" Phd. dissertation. University of Missouri. Columbia.

DELOITTE, HASKINS and SELLS (1979). Audit manual.

DODGE, H.F., ROMIG, H.G. (1944). "Sampling inspection tables" John Wiley (many subsequent editions).

EGGLESTON, R. (1978). "Evidence, proof and probability". Weidenfeld and Nicolson.

ELLIOTT, R.K. and ROGERS, J.R. (1972). Relating SS to audit objectives. Journal of Accountancy. July. pp. 46-55.

EVANS, D.H. (1980). "Audit fees - an intercompany comparison" MBA dissertation. University of Bradford. (Sup. by TWM)

FELIX, W.F. (1977) "Sampling risk v non-sampling risk" Auditing symposium 4. University of Kansas. pp. 47-56.

FELIX, W.F., GOODFELLOW, J.L. (1979) "Audit tests for internal control reliance." Symposium in Auditing Research. University of Illinois.

FIENBERG, S.E. NETER, J. and LEITCH, R.A. (1977). "Estimating the total overstatement error in accounting populations". Journal of the American Statistical Association. June.

FINLEY, D.R. (1978). "Controlling compliance testing with acceptance sampling." CPA Journal. Vol. 18. No. 2. December. pp. 30-35.

GARSTKA , S.J. (1977). "Computing upper error limits in dollar unit sampling" from Frontiers of auditing research. Ed. Cushing and Krogstad. University of Texas. pp. 163-204.

GOODFELLOW, J.L. LOEBBECKE, J.K. and NETER, J. (1974). "Some perspectives on CAV sampling plans". CA Magazine (Canada) October/November. pp. 46-53.

GREGORY, R.H. (1952). "The frequency and importance of errors in invoices received". Accounting Research. October Vol.3. No.4. pp. 332-339.

GUINN, R.E. (1973). "SS: a study and evaluation of its utilisation in auditing by CPA firms". P.hd dissertation. University of Alabama.

HEALY, R.E. (1964). "Sampling in auditing: a historical review." Price Waterhouse Review. Vol. 9. Winter. pp.30-41.

HERBERT, L. (1946) "Pratical sampling for auditors." Accounting Review. Vol. 21. No.4. October.

HILL, H.P. ROTH, J.L. and ARKIN, H. (1962). "Sampling in auditing" Ronald Press Co. New York

ICAEW (1979) List of members.

IJIRI, Y. and KAPLAN, R.S. (1971). "A model for integrating sampling objectives." Journal of Accounting Research. Spring 1971. pp. 73-87.

IJIRI, Y, LEITCH, R.A. (1980) "Steins paradox and audit sampling" Journal of Accounting Research. Vol. 18. No.1. pp 91-108.

JOHNSON, J.R., LEITCH, R.A. and NETER, J. (1979). "Characteristics of errors in accounts receivable and inventory audits". Unpublished paper. University of Georgia.

JONES, H.L. (1947). "Sampling plans for verifying clerical work". Industrial quality control. Vol. 3. Part 4. pp. 5-11.

JORDAN's Survey (1979). Britains quoted industrial companies.

JOSEPH, J.J. (1972). "Sampling and the independent auditor". Phd. dissertation. University of Oklahoma.

KAPLAN, R.S. (1973). "SS in auditing with auxiliary information estimators". Journal of Accounting Research" Autumn. pp. 238-258.

KAPLAN, R.S. (1975). "Sample size computations for Dollar Unit Sampling". Journal of Accounting Research. Vol. 13. pp. 126-133.

KECKER, F.M. (1955). "Admissibility in courts of economic data based on samples". Journal of Business. Vol. 28.

KINNEY, W.R. (1975). "Decision theory aspects of internal control system design". Journal of Accounting Research. Vol. 13. pp. 14-37.

KINNEY, W.R. (1977). "Integrating audit tests. Regression Analysis and DUS". CICA symposium. November. Paper 3.

KINNEY, W.R. (1978). "ARIMA and regression in analytical review". Accounting Review. January. pp. 48-60.

KRAFT, W.H. (1968). "SS for auditors - a new look". Journal of Accountancy. August 1968. pp. 49-56.

KRIENS, J. and DEKKERS, A.C. (1980). "SS in auditing" Stenfert Kroese. Leiden, Netherlands. (Written in Dutch).

LAUDEMAN, M.A. (1976). "Sampling risks associated with the application of variable sampling in the audit environment." Phd. dissertation. University of Akansas.

LESLIE, D.A. TEITLEBAUM, A.D. ANDERSON, R.J. (1980). "Dollar unit sampling". Pitman, London.

LOEBBECKE, J.K. NETER, J. (1975) "Considerations in choosing statistical sampling procedures in auditing." Journal of Accounting Research. Vol. 13. (Supplement) pp. 38-69.

MAGRUDER, E.T. (1950). "Statistical methods for appraising telephone property." C & P Telephone Co. Baltimore City.

MAXIM, L.D., CULLEN, D.E., COOK. F.X. (1976). "Optimal acceptance sampling plans for auditing batched stop and go v conventional single stage attributes plans." Accounting Review. January pp. 97-109.

McCRAY, J.H. (1980) "Tables for MEST bound overstatements and understatements". College of William and Mary. Williamsburg. Va.

McCRAY, J.H. (1980). "DUS. A procedure for calculating the maximum alpha and beta risks". College of William and Mary. Williamsburg, Virginia. USA. September.

McRAE, T.W. (1974). "Statistical sampling for audit and control." John Wiley. London.

McRAE, T.W. (1980). "The uses of SS in external auditing" - interim report." Presented to ICAEW - Audit standards Committee.

MEIKLE, G.R. (1972). "SS in an audit context". CICA. Toronto.

MONTEVERDE, R.J. (1955). "Some notes of reservation on the use of sampling tables in auditing." Accounting Review. Vol. 30. pp. 582-591.

MOTTERSHEAD, A.D. (1980). "Monetary unit sampling". Unpublished paper. (Josolyne, Leyton-Bennett).

NETER, J. (1949). "An investigation of the usefulness of SS methods in auditing." Journal of Accountancy. Vo. 87. No. 5. pp. 390-398.

NETER, J. (1952). "Some applications of statistics for auditing". Journal of the American Statistical Association. Vo. 47. No. 257. pp. 6-24.

NETER, J. (1952). "Sampling tables, an important statistical tool for auditors." Accounting Review. Vol. 27. No.4. pp. 475-483.

NETER, J. and LOEBBECKE, J.K. (1975). "Behaviour of major statistical estimators in sampling accounting populations." AICPA. New York.

NETER, J. LEITCH, R.A. FIENBERG, S.E. (1978). "Dollar unit sampling: multinomial bounds for total overstatement and understatements errors" Accounting Review. January.

NEWMAN, M.S. (1976). "Financial accounting estimates through SS by computer". John Wiley. New York.

NIGRA, A.L. (1963). "SS with variables". Internal Auditor. Summer. pp. 25-37.

PENNINGTON, R.R. (1979). "Company law" Butterworths. London.

PRYTHERCH, R.H. (1942). "How much test checking is enough?" Journal of Accountancy. Vol. 74. No.6. pp. 525-530.

PUSHKIN, A.B. (1978). "An investigation of the validity of auditing procedures used in mean-per-unit sampling plans." Phd. dissertation. Virginia Polytechnic.

RAJ, D. (1968). "Sampling theory" McGraw Hill. New York.

RAMAGE, J.G. KRIEGER, A.M. SPERO, L.L. (1979). "An empirical study of error characteristics in audit populations" Journal of Accounting Research. Vol. 17 (Sup) pp 72-113.

RANDALL, B., FRISHKOFF, P. (1976). "An examination of the status of probability sampling in the courts." Auditing Symposium 3, University of Kansas. (Editor H. Stettler) pp. 93-106.

RENEAU, J.H. (1978). "CAV bounds in dollar unit sampling : some simulation results." Accounting Review. Vol.1.53. July. pp. 669-680.

RESTALL, L.J., CZAJKOWSKI, P.J.(1969). "Computation of LIFO index. An SS approach". Management Accounting (US) September pp. 43-48.

ROBERTS, D.M. (1976). "A proposed sequential sampling plan for compliance testing". Symposium in auditing research. University of Illinois. pp. 159/168.

ROBERTS, D.M. (1978). "Statistical auditing". AICPA. New York.

ROBERTS, D.M. (1979). "Controlling audit risk". Faculty working paper. College of Commerce. University of Illinois. 1979.

ROSANDER, A.C., BLYTHE, R.H. (1951). "Sampling 1949 Corporate Income Tax returns". Journal of the American Statistical Association. June. pp. 233-241.

ROSTRON, R.Z. (1968). "SS as a useful audit tool" Haskins Sells selected papers. pp.211-216.

SAWYER, L.B. (1967). "The lawyer, the statistician and the internal auditor". Internal Auditor. Summer pp.9-18.

SMITH, T.M.F. (1976). "Statistical sampling for accountants" Accountancy Age books.

SMITH, K.A. (1972). "The relationship of internal control evaluation and audit sample size". Accounting Review April pp. 260-269.

SMURTHWAITE, J. (1965) "SS techniques as an audit tool" Accountancy. Vol. 76. No. 859. March. pp. 201-209.

SORENSEN, J.E. (1969). "Bayesian analysis in auditing" Accountancy Review. Vol. 44. No. 3. July. pp. 555-561.

SPROWLS, R.C. (1957). "The admissibility of sample data into a court of law". UCCA Law Review. Vol. 4.

STEPHAN, F.F. (1960). "Faculty advice about statistical sampling." Accounting Review. January.

STEPHAN, F.F. (1963). "Some statistical problems involved in auditing and inspection." Proceedings of the Business and Economic Statistics section, American Statistical Association, 1963. p.404. (Abstract only).

STRINGER, K. (1963). "Pratical aspects of SS in auditing." Proceedings of the Economics and statistics section of the American Statistical Association. 1963. pp. 405-411.

STRINGER, K. (1975). "A statistical technique for analytical review." Journal of Accounting Research (Sup) Vol. 13 pp. 1-13.

TAYLOR, R.G. (1974). "Error analysis in audit tests" Journal of Accountancy. May. pp. 78-82.

TEITELBAUM, L.L. BURTON, N.L. (1960). "Use of statistical techniques by the government." Journal of Accountancy. November. pp. 24-29.

TEITLEBAUM, A.D. (1973). "Dollar unit sampling in auditing." Proceedings of the American Statistical Association. 1973.

TEITLEBAUM, A.D., LESLIE, D.A. and ANDERSON, R.J. (1975). "An analysis of recent commentary in dollar unit sampling in auditing." Clarkson Gordon. Toronto. 1975.

TEITLEBAUM, A.D. McCRAY, J. and LESLIE, D.A. (1978). "Approaches to evaluating dollar unit samples." Proceedings of annual meeting of American Accounting Association.

TEITLEBAUM, A.D., ROBINSON, C.F. (1975). "The real risks in audit sampling." Journal of Accounting Research. Vol. 13. (Supplement) pp. 70-97.

TRACY, J.A. (1969). "Bayesian statistical confidence intervals for auditors." Journal of Accountancy. July. pp. 41-71.

TRUEBLOOD, R.M. CYERT, R.M. (1957). "Sampling techniques in accounting" Prentice Hall, Englewood Cliffs, N.J.

TUMMINS, M. DAVIS, E. (1979). "Block or random sampling?" Internal Auditor. Vol. 27. pp 57-64 May.

VANECEK, M.T. (1978) "Bayesian dollar unit sampling in auditing" Phd. dissertation. University of Texas at Austin.

VANCE, L.L. (1947) "Statistical sampling theory and auditing procedure". Proceedings of the Pacific Coast Economic Association.

VANCE, L.L. (1949) "Auditing uses of probabilities in selecting and interpreting test checks" Journal of Accountancy Vol.88 No. 3. September.

VANCE, L.L., NETER, J. (1956). "Statistical sampling for auditors and accountants" John Wiley, New York.

VAN HEERDEN, A. (1961). "Statistical sampling as a means of auditing" (in Dutch). Maanblad voor Accountancy en Bedrijfshuishoudkunde. 11. p.453.

WIGMORE, J.H. (1940). "The Anglo-American system of evidence". Little Brown & Co.

WILBURN, A. (1968) "Stop or Go sampling and how it works." Internal Auditor. May.

Some additional articles.

CORLESS, J.C. (1972) "Assessing prior distributions for applying Bayesian statistics in auditing". The Accounting Review. July. pp. 556-566.

CROSBY M.A. (1979) "Bayesian statistics in auditing, a comparison of probability elicitation techniques.".

Paper No. 702 - Krannert Graduate School of Management. Purdue University. (August).

CYERT, R.M. and others (1960) "SS in the audit of Air Force Motor vehicle inventory".

Accounting Review Vol. 36. pp 567-673.

FELIX, W.L. (1976) "Evidence on alternative means of assessing prior probability distributions for audit decision making. The Accounting Review, October pp. 880-807.

SNEED, F.R. (1979) "A study of the effects of conservatism on the evidential sample size decisions made by auditors.".

Phd. dissertation. North Texas State University.

McCRAY, J.H. (1981) "Dollar Unit Sampling: a model for calculating the upper bound".

College of William and Mary, Williamsbury, Va U.S.A. Unpublished.

Alpha Risk.

c	.01	.05	.10	.25	.50
0	.010	.051	.105	.288	.693
1	.149	.355	.532	.961	1.68
2	.436	.818	1.10	1.73	2.67
3	.823	1.37	1.75	2.54	3.67
4	1.28	1.97	2.43	3.37	4.67
5	1.79	2.61	3.15	4.22	5.67

Beta Risk.

c	.01	.05	.07	.10	.15	.20	.25	.30	.40	.50
0	4.61	3.00	2.66	2.30	1.90	1.61	1.39	1.21	.92	.69
1	6.64	4.74	4.34	3.89	3.38	3.00	2.69	2.44	2.02	1.68
2	8.41	6.30	5.83	5.32	4.73	4.28	3.92	3.62	3.11	2.67
3	10.0	7.76	7.25	6.68	6.02	5.52	5.11	4.76	3.18	3.67
4	11.6	9.15	8.88	7.99	7.3	6.72	6.27	5.89	4.24	4.67
5	13.1	10.5	10.21	9.27	8.5	7.91	7.42	7.01	6.29	5.67

x = acceptable error rate. The auditor does not want to reject populations with an error rate as low as this.

u = unnacceptable error rate. The auditor wishes to reject populations with an error rate as high as this.

a = alpha risk. The risk of rejecting populations with an error rate of x% or less.

b = beta risk. The risk of accepting populations with an error rate of u% or above.

c = critical number of errors discovered in audit sample.

s = audit sample size.

Exhibit A.1. KAPLAN'S TABLE for calculating MUS sample size.This table allows the auditor to calculate the required audit sample size taking into account both alpha and beta risk. The table thus allows the auditor to use an acceptance sampling approach to MUS sampling.
Source: Kaplan(1975)

Appendix A.1. (Continued)

Using Kaplans table.

(Kaplans method is very conservative since it assumes before the sample is drawn that all errors to be discovered will be 100% tainted. We noted in Chapter Three that most overstatement errors are tainted by less than 10%).

The method.

1. Decide on x, u, a and b.
2. Select alpha risk factor and beta risk factor from tables for c = o.

   Let alpha risk factor = A

   Let beta risk factor = B
3. DivideA by x = f.

   Divide B by u = g.
4. If f $\geqslant$ g round up g to next whole number and this gives us sample size s.
5. If f $<$ g select new alpha and beta risk factors for c = 1
6. If f $\geqslant$ g go back to (4).

   If f $<$ g add 1 to c and repeat until f $\geqslant$ g.
7. Draw MUS sample size s, audit sample in normal way, if number of 100% errors discovered exceeds c reject population, otherwise accept.

   Important note: If errors are tainted less than 100% add the taintings together to make up "equivalent 100% errors". For example.

   1 x 50% tainted error.

   2 x 20% tainted errors.

   1 x 10% tainted error.

equal 1 x 100% tainted error.

(See Kaplan (1975) p 133).

Example.

1. Let $x = 0.5\%$ (i.e. 0.005)
$u = 5\%$ (i.e. 0.05)
$a = 25\%$
$b = 5\%$

2. Calculation

Alpha Risk				Beta Risk		
c	x	a (.25)	$f = \frac{a}{x}$	u	b (.05)	$g = \frac{b}{u}$
0	.005	0.288	57.6	.04	3.00	75
1	.005	0.961	192.2	.04	4.74	118.5

3. For c = 1, f > g, therefore required MUS sample size is 119 and the auditor rejects on <u>more than</u> one 100% error.

4. Suppose the following errors are found in the audit sample.

1 x 5% tainting
1 x 20% tainting
<u>1</u> x <u>50%</u> tainting
<u>3</u> <u>75%</u>

5. Although 3 errors are found their worth is only equal to 75% of a full 100% error. Therefore the population is accepted.

LIST OF EXHIBITS

LIST OF EXHIBITS (continued)

LIST OF EXHIBITS (continued)

INDEX

INDEX (continued)

INDEX (continued)

INDEX (continued)

INDEX (continued)

INDEX (continued)

AUTHOR INDEX